The Open University

CH00482427

Block A

Computer algebra

About this course

MS325 Computer algebra, chaos and simulations uses the software package *Maple*™ (copyright Maplesoft™, a division of Waterloo Maple Inc, 2007) which is provided as part of the course. Maple is a computer-assisted algebra package and its usage is the main subject of *Block A Computer algebra*. Advice on such matters as the installation of Maple, the loading and saving of Maple worksheets and other basic 'getting-started' issues is covered in the *Computing Guide*.

Maplesoft™ and Maple™ are trademarks of Waterloo Maple Inc. All other trademarks are the property of their respective owners.

The cover image is composed of a photograph of the Saturnian moon Hyperion (courtesy of NASA) overlaid with a time-T map (in yellow). The view of Hyperion was taken during the close flyby of the spacecraft Cassini on 26 September 2005. Time-T maps are covered in *Block B Chaos and modern dynamics*. This one shows Hyperion's spin rate plotted against its orientation and was generated using Maple from a mathematical model. Regions containing an apparently random scatter of dots indicate chaotic motion in Hyperion's spin angle, the so-called chaotic tumbling.

This publication forms part of an Open University course. Details of this and other Open University courses can be obtained from the Student Registration and Enquiry Service, The Open University, PO Box 197, Milton Keynes, MK7 6BJ, United Kingdom: tel. +44 (0)870 300 6090, e-mail general-enquiries@open.ac.uk

Alternatively, you may visit the Open University website at http://www.open.ac.uk where you can learn more about the wide range of courses and packs offered at all levels by The Open University.

To purchase a selection of Open University course materials, visit http://www.ouw.co.uk, or contact Open University Worldwide, Michael Young Building, Walton Hall, Milton Keynes, MK7 6AA, United Kingdom, for a brochure: tel. +44 (0)1908 858793, fax +44 (0)1908 858787, e-mail ouw-customer-services@open.ac.uk

The Open University, Walton Hall, Milton Keynes, MK7 6AA.

First published 2008.

Edited, designed and typeset by The Open University, using the Open University TEX System.

Printed and bound in the United Kingdom by Hobbs the Printers Limited, Brunel Road, Totton, Hampshire SO40 3WX.

ISBN 978 0 7492 1586 6

1.1

Contents

UNIT 1 Basic commands

Study guide

There are 9 sections in this unit, which are intended to be studied in order. Each section contains worked examples and exercises, and you should make sure that you work through all of these. In particular, several exercises contain new material that you are asked to explore for yourself.

We recommend that you divide your study time into sessions, each lasting roughly three hours. The pattern of study for each three-hour session might be as follows.

> Study session 1: Section 1.1
> Study session 2: Sections 1.2 and 1.3
> Study session 3: Section 1.4
> Study session 4: Section 1.5
> Study session 5: Section 1.6
> Study session 6: Section 1.7
> Study sessions 7 and 8: Section 1.8
> Study sessions 9 and 10: Section 1.9

Before studying this unit, you should have read the course *Computing Guide*. You are not expected to have any prior knowledge of Maple, although you should have installed it on your computer, as described in the *Computing Guide*. Regular access to a computer is required in order to work through this unit.

Introduction

Maple is one of a number of algebraic computing systems. It can be used as a basic calculator, for instance to evaluate $\ln 3$ or $\sqrt{2}$ to any number of decimal places (within reason), but its great strength is its ability to manipulate *algebraic* expressions. For example, it can be used to solve equations exactly, manipulate symbolic matrices and vectors, factorize polynomials and simplify rational expressions. It knows about complex numbers, and it can also perform many of the standard operations of analysis such as evaluating limits, derivatives and integrals. It has good two- and three-dimensional graphics facilities so, for example, solutions of ordinary differential equations can be found and plotted. It is also a programming language, and Maple programs enable quite complicated operations to be carried out. Indeed, most of Maple itself is written in Maple!

This unit introduces you to several of these features and sets the scene for many more. First, you will learn how to input simple expressions into Maple. You will then see how to plot simple two-dimensional line graphs and how to solve certain equations both symbolically and numerically. The unit then introduces some of the basic calculus facilities of Maple, so that you will be able to differentiate and integrate algebraic expressions. You will also learn about evaluating sums and products. Finally, we demonstrate several ways of manipulating Maple's output into required and usable forms.

You may well have some experience already of a computer algebra system. This course introduces you to Maple because it is a good all-round package that suits the needs of professional mathematicians and physicists very well. Its language and structure enable calculations to be automated, and at this stage in your studies you should have the necessary mathematical knowledge to make good use of this.

For example, if you have studied the course MST209 then you will have used Mathcad extensively.

Although this unit appears to introduce a large number of Maple commands, many of these, such as `plot`, `sum`, `diff` and `int`, are almost self-explanatory. In any case, a list of commands, giving more information, is contained in the glossary of the *Course Handbook*, and you are encouraged also to use the Maple 'help' facility as explained in the *Computing Guide*. Listings of the Maple code contained in this unit can be found on the course website and on the DVD sent with the course materials. In order to become technically adept at applying Maple, it is important that you attempt *all* the exercises contained here; it is not sufficient simply to read the solutions without attempting the questions first. Remember that Maple is best learned by actually *using* it to *do* mathematics, and this should be practised as often as possible.

1.1   *Introduction to Maple*

In this first section we give a brief introduction to Maple, outlining some of its basic features and illustrating them with some short examples. It is assumed that you have read the *Computing Guide*, have installed Maple, and have opened a 'Classic Worksheet'. You will need to type in all the commands that you see in this and subsequent sections; in this way you will become increasingly familiar with the basic ideas.

1.1.1   *Numbers and constants*

First, note that in the Maple Classic Worksheet the **prompt** is the symbol >. Commands are entered to the right of the prompt, and each command ends with either a colon or a semicolon. If a colon is used then the command is executed but the output does not appear on the screen, whereas when a semicolon is used the output appears on the screen. Once you have finished entering a command line, you should press the Return/Enter key. To illustrate this, try the following. At the prompt,

The colon is sometimes called the **silent terminator** and the semicolon the **noisy terminator**.

type `12/15;` then press the Return/Enter key. You should see the following:

> `12/15;`

$$\frac{4}{5}$$

Notice how Maple transforms the input by reducing the fraction to its lowest terms.

However, typing

> `12/15:`

suppresses the output so that no output is seen (this is useful, for example, if the output is extensive and not specifically required). If you forget the colon/semicolon then Maple will issue a warning, and the easiest thing to do then is to move the cursor back to the end of the line just typed (using the mouse or the arrow keys) and add the colon/semicolon. Pressing Return/Enter should then enable the command to be executed, and will remove the warning; there is no need to delete it. To see some more commands, try the following:

> `3+18;`

$$21$$

> `40*21;`

$$840$$

> `3!;`

$$6$$

In the next example, you should press the Return/Enter key only after typing in the whole line:

> `2^3; exp(2); sin(Pi); cos(3*Pi/4);`

$$8$$
$$e^2$$
$$0$$
$$-\frac{\sqrt{2}}{2}$$

The final example above illustrates some important features. First, note that several commands can be put on a single line provided that each ends with its own semicolon (or colon, if preferred). Second, Maple is case-sensitive, and `Pi` with a capital `P` represents the number $3.141\,592\,653\ldots$ (more of this in a moment). Finally, note that multiplication must be made explicit with the $*$ symbol, so that 3π is input as `3*Pi`; powers are input via the `^` symbol, as shown for 2^3 above.

You should also observe in the above examples that Maple uses exact arithmetic, giving exact answers wherever possible (so that $\sqrt{2}$ is not evaluated to a decimal number, for example). If we want decimal approximations, then in certain circumstances we can force Maple to produce them by including at least one decimal point in a numerical expression in the input, as illustrated in the following examples:

> `sqrt(2.0);`

$$1.414213562$$

> `22/7.0;`

$$3.142857143$$

These decimal numbers are given to ten significant figures, which is the default setting in Maple. You will see shortly how to alter this.

If the decimal point follows an integer, then it does not need to be followed by 0, so (for example) the first input above could equally well be:

> `sqrt(2.);`

$$1.414213562$$

However, note that including a decimal point in the input does not *guarantee* a decimal output (try typing `3.0*Pi` at the prompt, for example). In Subsection 1.2.2 you will be introduced to the `evalf` command, which *will* always return an output in decimal form.

To check your understanding of the material covered so far, try the following exercise.

Exercise 1.1

(a) Use Maple to evaluate the following exactly:

$$19 \times 99, \quad 3^{20}, \quad 25!, \quad \tfrac{3}{13} - \tfrac{26}{27}, \quad 2^{(2^2)}.$$

(b) Now evaluate the following to ten significant figures (and remember that you may need to include a decimal point in the input):

$$\tfrac{21}{23}, \quad 17^{1/4}, \quad \tfrac{1}{99!}, \quad 0.216^{100}.$$

As you saw above with `Pi`, Maple has some names reserved for **constants**. The list of these is short, and the ones that you should be aware of now are shown in Table 1.1 below.

Table 1.1   Some of Maple's constants

`Pi`	$3.1415\ldots$
`I`	$\sqrt{-1}$
`infinity`	∞

Missing from this table, because there is no special symbol for it, is the number e, that is, $\exp(1)$ or $2.7182\ldots$. You should not be tempted to input e or E, as Maple will not recognize either of these. However, typing `exp(1)` at the prompt gives the output

```
>   exp(1);
```
$$\mathbf{e}$$

where the **e** produced by Maple does indeed mean $2.7182\ldots$ (try typing `exp(1.0)` at the prompt to see what happens). Within Maple, you should be careful not to confuse the **e** in bold above, which when output denotes $\exp(1)$, with the letter e in italics which has no particular meaning.

Now work through the following exercise, which uses the constant `I`.

Exercise 1.2

As given in Table 1.1, Maple denotes the square root of -1 by `I`. Suppose that we wish to express $(2+3i)^4$ in the form $a+bi$, where a and b are real. The easiest way of achieving this is by typing

> This is more usually denoted by i.

```
>   (2+3*I)^4;
```
$$-119 - 120\,I$$

Express the following in the form $a+bi$:

$$(5-8i)^2, \quad i^{-1}, \quad \frac{1+i}{5+2i}, \quad \left(\frac{4}{3}+\frac{2i}{5}\right)^4.$$

1.1.2 Variables and names

Variables in Maple are denoted by sequences of letters, digits and underscores (with the exception that a variable may not begin with a digit). For example

Remember that Maple is case-sensitive, so a1 and A1 are different variables.

    x, f, a1, B2, initial_conditions

are all legitimate variables. There are a few words that are reserved by Maple and so cannot be used for variable names. You have already seen Pi, which is one, and for and while are others, because these are needed for certain Maple operations (many of which you will learn about throughout this course). For a list of reserved names, type ?reserved at the prompt.

Variables can be either **free** or **assigned**. A free variable is one which has no value associated with it, whereas an assigned variable takes a particular value (which might be a number, or a more complicated expression). To assign a value to a variable, use the **assignment operator :=**, that is, a colon followed by an equals sign. Hence A := B (followed by the usual colon or semicolon and then pressing the Return/Enter key) will assign to A the value of B. So typing

Another word for 'free' in this context is 'unbound', and another word for 'assigned' is 'bound'.

> a := 20;
$$a := 20$$

If you have studied the course MST209, then you will already have come across := as an assignment operator, when using Mathcad. Note that, unlike in Mathcad, in Maple you need to type both the colon : and the equals sign = in order to obtain :=.

assigns the value 20 to the variable a. (There is no need to put a space either side of := when typing a := 20, but it makes the input slightly clearer if you do so.) This can be checked by asking Maple what value a takes, which is done simply by typing a followed by a semicolon, as illustrated here:

> a;
$$20$$

Note that there is a direction implicit in assignment; typing 20 := a does *not* assign the value 20 to the variable a.

Extending this example, assign b the value a+3 as follows:

> b := a+3;
$$b := 23$$

Maple already knows that a has the value 20 and so has evaluated a+3 directly as 23.

Assignments can be overwritten and Maple will remember only the most recent one that it has been asked to execute. So b, which above took the value a+3, can now be assigned the value 10 (say) as follows:

> b := 10:
> b;
$$10$$

Warning

A frequent mistake in Maple is to attempt to use the equals sign = for assignment purposes, rather than the assignment operator :=. It is essential to use := when assigning an expression to a name.

Uses of = in Maple will be illustrated later in this unit.

The following illustration of assignment concerns the reserved variable Digits, which controls the number of digits displayed as output and used

The capital D is important here; digits is just a free variable.

in computation. The default value of `Digits` is 10, but this can be changed using the assignment operator, as shown here:

```
>  22/7.0;
```

$$3.142857143$$

```
>  Digits;
```

$$10$$

```
>  Digits := 20;
```

$$Digits := 20$$

```
>  22/7.0;
```

$$3.1428571428571428571$$

To check your understanding of assigning variables, try the following exercises.

Exercise 1.3

Assign to a the value 10, to b the value 20, and to c the value a+2*b. Then use Maple to evaluate a+b+c.

Exercise 1.4

All rational numbers p/q, where p and q are integers ($q \neq 0$), have a terminating or (eventually) repeating decimal form. By increasing the number of `Digits` as appropriate, find the exact decimal form of $21/23$.

Exercise 1.5

Consider the following series of commands. After inputting them all as shown, what would the final line (that is, `z;`) give as output? (Try to work this out *before* inputting the commands yourself!)

```
>  a := 3; b := 2;
```

$$a := 3$$
$$b := 2$$

```
>  # In the following line the output is deliberately
   # suppressed
>  z := a^2+b^2+x^2:
>  a := 4;
```

$$a := 4$$

```
>  z;
```

Exercise 1.5 illustrates the use of one other new symbol, **#**. Maple will ignore everything after this symbol, until the start of the next line. It helps to keep track of things if you put in lines of explanation like this from time to time; this will become particularly important when writing the longer procedures that will be encountered later in the course.

You may be wondering how to remove assignments to variable names. The easiest way to do this is to type `restart:` or `restart;` (both have the same effect) at the prompt; this will reset everything and enable you to start again from scratch. This is always an option if the expected output is not forthcoming; going back to the start of a sequence of command lines and adding in `restart:` before the first command might work wonders.

`restart:` will reset `Digits` to 10, for example.

Indeed, it is strongly recommended that you start all exercises and worksheets with `restart:`. However, do bear in mind that `restart:` removes the effect of *all* previous input. More subtle ways of unassigning names – and in particular how to unassign a single name without disturbing others – will be covered later in this unit.

1.1.3 Elementary functions

Maple has many built-in functions, most of which are fairly obvious. Several have already been introduced, for example:

```
>  cos(3*Pi/4);  sqrt(2.0);  exp(5);  sqrt(2);
```

$$-\frac{\sqrt{2}}{2}$$
$$1.414213562$$
$$e^5$$
$$\sqrt{2}$$

Note that, as already seen, where possible Maple will produce exact answers unless you request an output value in decimal form by including a decimal point in the input.

For the moment, the functions that you should be familiar with are those in Table 1.2 below. (In each case the function needs to be followed by an argument in brackets, as illustrated in the examples above.)

Remember that help with any of these may be found by typing `?` followed by the particular function you are interested in; for example, `?abs` will give you more information on the `abs` function.

Table 1.2   Some of Maple's built-in functions

cos	cosine
sin	sine
tan	tangent
sec	secant
csc	cosecant
cot	cotangent
exp	exponential
ln or log	natural logarithm
log10	logarithm to base 10
abs	absolute value
sqrt	square root

To obtain a logarithm to any other base, enclose the base in square brackets; for example, `log[8](64)` will be evaluated as 2.

Before you practise these, here are some more examples.

```
>  (sin(Pi/4))^2;  sin(Pi/4)^2;  sin(3)^2;
```

$$\frac{1}{2}$$
$$\frac{1}{2}$$
$$\sin(3)^2$$

```
>  exp(2);  ln(%);
```

$$e^2$$
$$2$$

```
>  abs ( -3);
```

$$3$$

In the above examples you should note the following.

- Maple's notation for $\sin^2 x$ (and indeed $\sin^3 x$, $\cos^2 x$, etc.) is potentially misleading. When Maple outputs $\sin(3)^2$ in the example above, it means $\sin^2 3$, that is, $(\sin 3)^2$, and not $\sin(3^2)$. Similarly, `sin(x)^2` or `(sin(x))^2` is used to input $\sin^2 x$; `sin^2x` would not be accepted as input.

- The symbol %, known as the **ditto operator**, represents 'the value of the expression most recently evaluated'. You should use it with great care but it can be helpful. It is better practice, however, to assign the output a name, and then to use this name in subsequent work, as illustrated by the following: `ans := exp(2); ln(ans);`.

- Within reason, Maple ignores the presence of spaces, as illustrated in the final example above. However, it would not have evaluated `ab s(-3)`, for example.

Now try the following exercises.

Exercise 1.6

(a) Use Maple to evaluate the following:
$$\tan(\pi/3), \quad \cos^2(\pi/3), \quad \sec(\pi/6), \quad 1 + \cot^2(\pi/4), \quad \operatorname{cosec}^2(\pi/4).$$

(b) Evaluate the following, giving numerical answers to ten significant figures where appropriate:
$$\sqrt{3 + \sqrt{3 + \sqrt{3}}}, \quad e^{3\ln 4}.$$

Exercise 1.7

There is a Maple function, namely `ithprime(k)`, that returns the kth prime number. By using the Maple help facility, if necessary, investigate this function and use it to find the 100th, 1000th and 10 000th primes.

Exercise 1.8

(a) Use the Maple help facility to find out how to evaluate $\arcsin(1/2)$.

(b) Evaluate $\sec(\arctan x)$ in terms of x.

That ends this first section on Maple. Before proceeding any further, you should take stock and make sure that you are happy with everything covered so far. In particular, you should feel confident using the material in Tables 1.1 and 1.2 as well as the following: `:`, `;`, `:=`, `restart`, `%`, `#` and `Digits`. The rest of this unit, and indeed the whole course, will build on these, so it is essential that you understand them.

1.2   *The* eval *family*

In this section, three of the *evaluation* commands that make up the eval family are introduced. The material here is fairly brief; you will be given plenty of opportunity to practise the commands throughout the rest of this course.

1.2.1   eval

The command eval is used to **eval**uate expressions. For example, suppose we want to evaluate the expression

```
>   y := x^2+3*x-2;
```

$$y := x^2 + 3\,x - 2$$

at $x = 1$. This is achieved by using the command

```
>   eval(y,x=1);
```

$$2$$

What is useful about this is that y itself has not changed, as verified by typing

```
>   y;
```

$$x^2 + 3\,x - 2$$

so if we now wish to evaluate y at a different value of x then this is easily done.

To practise inputting expressions and evaluating them, try the following exercise.

Exercise 1.9

In each of the following, evaluate the expression at the given value.

(a) $x^3 - 3x^2 + 2x - 1$, at $x = 5$

(b) $\sin x \cos^3 x$, at $x = \pi/4$

(c) e^t, at $t = 1$

(d) $\ln\!\left(u + 1 + \sqrt{u^2 - 3}\right)$, at $u = 2$

(e) $3\sin(2t) + 4\cos(3t)$, at $t = \pi/3$

To evaluate an expression containing more than one variable, put the variable values in curly brackets, separated by commas. As an example, suppose that we wish to evaluate $r = x^2 + y^2 + z^2$ at the point $(x, y, z) = (1, 2, 3)$. This is achieved by typing

The curly brackets denote a **set**; you will learn more about sets in *Unit 2*.

```
>   r := x^2+y^2+z^2;
```

$$r := x^2 + y^2 + z^2$$

```
>   eval(r,{x=1,y=2,z=3});
```

$$14$$

to give $r = 14$.

1.2.2 `evalf`

The `evalf` command (**eval**uate using **f**loating-point arithmetic) is used to evaluate an expression to a **floating-point** number, that is, a number rounded to a certain number of significant figures. The default number of significant figures used is 10 (the default value of `Digits`). This may be changed either by re-assigning `Digits` – which will then produce a global change in all subsequent expressions until it is re-assigned again – or by using an additional optional argument, enclosed in square brackets:

> You saw a re-assignment of `Digits` in Subsection 1.1.2.

```
>   evalf(sqrt(2));
```
$$1.414213562$$

```
>   Digits;
```
$$10$$

```
>   Digits := 20:     evalf(sqrt(2));
```
$$1.4142135623730950488$$

```
>   evalf[7](sqrt(2));
```
$$1.414214$$

The [7] placed immediately after `evalf` indicates that `sqrt(2)` is to be evaluated to 7 significant figures; note how `evalf` has rounded conventionally, so that the seventh significant figure, 3, has been rounded to 4. An advantage of putting the required number of significant figures in square brackets in this way is that the value of `Digits` is unchanged.

Care is needed when numerically evaluating very complicated expressions, as severe cancellation may result in the numerical accuracy of a computation being compromised. For example, the difference $\sin(2 \times 10^k \pi + y) - \sin y$ is identically zero for any non-negative integer k, but when $2 \times 10^k \pi + y$ is evaluated using `evalf`, a non-zero answer is returned for the whole expression:

```
>   y := 0.2:     k := 5:
>   sin(evalf[4](2*10^k*Pi+y))-sin(y);
```
$$-0.4092089823$$

```
>   sin(evalf[10](2*10^k*Pi+y))-sin(y);
```
$$0.0000804053$$

The first result is not accurate to any decimal places at all, while the second is accurate to only four! This is because $200\,000\pi + 0.2$ evaluated to four digits is in *absolute* terms rather a long way from its true value. While increasing the number of digits usually improves numerical accuracy, as illustrated above, sometimes it is better not to evaluate numerically at all:

```
>   y := 0.2:     k := 5:
>   sin(2*10^k*Pi+y)-sin(y);
```
$$0.$$

Exercise 1.10

The following two expressions are both approximations to π that were discovered by the Indian mathematician Srinivasa Ramanujan (1887–1920):

$$\pi_1 = \frac{12}{\sqrt{190}} \ln\left((2\sqrt{2} + \sqrt{10})(3 + \sqrt{10})\right)$$

and

$$\pi_2 = \sqrt{\sqrt{9^2 + \frac{19^2}{22}}}.$$

Use the `evalf` command to find the absolute errors $|\pi_1 - \pi|$ and $|\pi_2 - \pi|$. Hence determine how good these approximations actually are.

1.2.3 `evalc`

This subsection introduces the command `evalc` (**eval**uate using **c**omplex number arithmetic). This assumes that all unknown variables in an expression are real and puts a complex number in its canonical form $a + bi$, where a and b are real.

To set the scene, suppose that we wish to express $\dfrac{3 + 4i}{5 + 6i}$ in the form $a + bi$; this is achieved by typing

> You saw examples similar to this in Exercise 1.2.

```
>  (3+4*I)/(5+6*I);
```

$$\frac{39}{61} + \frac{2}{61}I$$

Without the need for `evalc`, Maple has shown that $\dfrac{3 + 4i}{5 + 6i} = \dfrac{39}{61} + \dfrac{2}{61}i$.

However, in certain circumstances evaluation to the form $a + bi$ has to be forced using `evalc`. This is illustrated in the following example, where it is required to express $\ln(3 + 4i)$ in the form $a + bi$:

> The natural logarithm of a complex number z is defined by $\ln z = \ln|z| + i \arg(z)$, where $|z|$ is the complex modulus and $\arg(z)$ is the complex argument.

```
>  z := 3+4*I;
```

$$z := 3 + 4\,I$$

```
>  ln(z);
```

$$\ln(3 + 4\,I)$$

This is not yet in the desired form, so type

```
>  evalc(ln(z));
```

$$\ln(5) + \arctan\left(\frac{4}{3}\right)I$$

to see that $\ln(3 + 4i) = \ln 5 + i \arctan(4/3)$.

`Re` and `Im` can be used to extract the real and imaginary parts, respectively, of an expression, as illustrated in the following example:

```
>  Re((1+3*I)^2);   Im((1+3*I)^2);
```

$$-8$$
$$6$$

In the previous simple example there was no need for `evalc`, but as above it can be used to force a required output when necessary:

```
>   z := cos(t)+I*sin(t);
```
$$z := \cos(t) + \sin(t)\, I$$

```
>   Re(z^4);
```
$$\Re((\cos(t) + \sin(t)\, I)^4)$$

In the above output, the real part of `z^4` has not been found. However, typing

```
>   evalc(Re(z^4));
```
$$\cos(t)^4 - 6\cos(t)^2\sin(t)^2 + \sin(t)^4$$

evaluates the real part of $(\cos t + i\sin t)^4$ as $\cos^4 t - 6\cos^2 t \sin^2 t + \sin^4 t$.

To check your understanding of `evalc`, try the following exercises.

> When using `evalc`, it is assumed that `t` is real, as it is an unknown variable. Maple does not usually make this assumption about unknown variables, but `evalc` is a special case.

Exercise 1.11

De Moivre's theorem states that
$$(\cos\theta + i\sin\theta)^n = \cos(n\theta) + i\sin(n\theta)$$
for all integers n. Use this theorem to find expressions for $\cos(2\theta)$ and $\sin(2\theta)$ in terms of $\cos\theta$ and $\sin\theta$.

> If you input the English word for a Greek letter, then Maple outputs the Greek letter itself. For example, try `delta:=1;` and `Delta:=1;`. (The constant `Pi` is an exception.)

Exercise 1.12

If $z = \frac{3}{2}\ln 2 + i\frac{\pi}{4}$, express e^z in the form $a + bi$.

1.3   *Simple line graphs with the* plot *command*

A useful feature of Maple is the ease with which many different types of graphs may be drawn, and throughout this course we shall make extensive use of Maple's graphics facilities. Here we introduce the simplest type of graph, namely the graph of a given function of a single real variable, x, on a given interval. Other types of plots will be described later in the course.

Most graphs are drawn with variants of the `plot` command, which has several forms, each with many optional arguments that determine the appearance of the resulting graph. The simplest of these forms, with a few of the more important options, is introduced in this section; if at any time you wish to see a list of available options, then type `?plot,options` at the prompt to access the relevant help pages.

The simplest graphs are those of functions of a single real variable; for example, the graph of the function $\sin(10/(1+x^2))$ for $0 \le x \le 10$ is given by the command

```
> plot(sin(10/(1+x^2)),x=0..10);
```

The '..' in x=0..10 is input by typing two successive full stops.

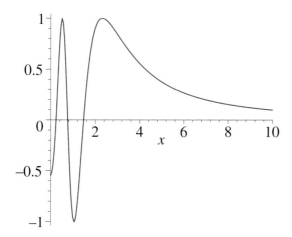

Figure 1.1   Graph of $\sin(10/(1+x^2))$ for $0 \le x \le 10$

This graph is shown in Figure 1.1; the appearance on your screen may be slightly different. The first argument in this command is the formula for the function, and the second is the **range** of x; if we needed the graph for $10 \le x \le 100$, then we would replace x=0..10 by x=10..100. On the other hand, the command plot(sin(10/(1+x^2)),x); would plot the graph over the default range $-10 \le x \le 10$, which is rarely of any use.

The word **range** here is a Maple term; it does not have the usual meaning of the range of a function in mathematics.

The range of the vertical axis can also be specified; if we want to see the values of $y = \sin(10/(1+x^2))$ only in the range $0 \le y \le 1$, then we use the command

```
> plot(sin(10/(1+x^2)),x=0..10,0..1);
```

which gives the graph shown in Figure 1.2.

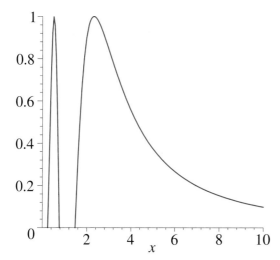

Figure 1.2   Graph of $\sin(10/(1+x^2))$ for $0 \le x \le 10$ and $0 \le y \le 1$

Exercise 1.13

Assign the expression $\sin(10\cos x)$ to the variable y, and use the command plot(y,x=-Pi..Pi); to draw the graph of this function for $-\pi \le x \le \pi$.

The colour of a graph is changed using an optional argument in the plot command. For example, the graph of $y = \sin(10\sin x)$ for $0 \le x \le \pi$ is drawn in blue with the command

```
>  plot(sin(10*sin(x)),x=0..Pi,colour=blue);
```

(Note that the order of these arguments is important; the range of x values must come before optional arguments such as `colour`.) For a full list of colours that are predefined in Maple type `?plot,color` at the prompt. (Note the use of American spelling here – typing `?plot,colour` will take you to a list of options from which you can choose `?plot,color`.)

One of the more useful features of the `plot` command is the ability to draw graphs of many functions on the same plot. Thus the graphs of $\sin x$ and $\cos x$ for $0 \le x \le \pi$ can be compared with the command

```
>  plot([sin(x),cos(x)],x=0..Pi,colour=[coral,turquoise]);
```

Here the two functions are inserted as the **list** `[sin(x),cos(x)]` – the enclosing square brackets are important. Also, we have used the colour option in the form `colour=[coral,turquoise]`, where again the square brackets are essential, and the colour order matches the function order. (It is not necessary to define colours when plotting two or more graphs, because Maple will automatically plot each curve in a different colour, but you may prefer to do so.) A legend can be added to the plot with the `legend` option, as in the following example; note here that not only is the order of the items in the legend important, as always, but use of the quotation marks is essential.

You will meet lists again in *Unit 2*.

```
>  plot([sin(x),cos(x)],x=0..Pi,colour=[red,gold],
      legend=["sin x","cos x"]);
```

Any number of graphs, within reason, can be combined in this manner simply by including more functions in the list. Thus the command

```
>  plot([x,x^2,x^3],x=0..1,colour=[red,blue,black]);
```

draws the graphs of $y = x$ (red), $y = x^2$ (blue) and $y = x^3$ (black) for $0 \le x \le 1$, all on the same plot.

Exercise 1.14

Draw the graphs of the five functions $y_1(x) = 1$, $y_2(x) = 1 + x$, $y_3(x) = 1 + x + x^2/2!$, $y_4(x) = 1 + x + x^2/2! + x^3/3!$ and $y_5(x) = \exp(x)$ on the same plot for $0 \le x \le 1$.

Simple changes to the presentation of graphs can be made after they have been drawn by placing the cursor in the vicinity of the figure and clicking the left mouse button. Then new buttons will appear on the third bar at the top of the worksheet to allow you to make some changes; you can see what these are by moving the cursor over each item in turn. What you may find particularly useful is the extreme left-hand box on the bar, which gives the coordinates of the cursor when you click on the plot; this is helpful for finding approximate coordinates of roots or stationary points. To practise this, try the following exercise.

Right-clicking on a figure gives a pop-up menu that is also useful for manipulating plots.

Exercise 1.15

Find the approximate values of the x-coordinates of all the real solutions of the nonlinear simultaneous equations

$$y = \sin x,$$
$$y = x^3 - 5x^2 + 4.$$

Sometimes it is necessary to plot graphs of functions that are undefined at particular points. For instance, $y = \tan x$ is unbounded at $x = (n + \frac{1}{2})\pi$, for integer n. The plot command will attempt to draw the graphs of such functions, but without limits on the y-coordinate the scale is normally distorted. Limiting the range of the y-coordinate overcomes this problem but may introduce unwanted vertical lines. For example, try the following commands:

```
>  plot(tan(x),x=0..7);
```

```
>  plot(tan(x),x=0..7,-10..10);
```

These problems are solved with the use of the optional argument `discont=true`. Thus the command

```
>  plot(tan(x),x=0..7,-10..10,discont=true);
```

gives the graph shown in Figure 1.3.

This command tells Maple that the function being plotted has at least one discontinuity in the range of the plot.

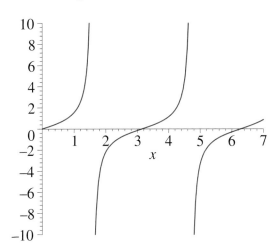

Figure 1.3   Graph of $\tan x$ using the `discont=true` option

Exercise 1.16

Plot the graphs of $y = \tan x$ and $y = 1/x$ on the same diagram over the range $0 \le x \le 15$, using the `discont=true` option. Use your diagram to find the first five positive roots of the equation $x \tan x = 1$ correct to one decimal place.

Exercises 1.15 and 1.16 will be revisited in Subsection 1.4.2, where Maple will be used to find better approximations to the roots.

That completes this very brief introduction to Maple's graphics facilities. As mentioned earlier, you will extend your knowledge of the various plot commands and options throughout the course, but in the meantime we move on to look at two members of the `solve` family of commands.

1.4 The solve *family*

The main aim of this section is to introduce `solve` and `fsolve`, two members of the `solve` family of commands. You will also revise `eval`, and learn about the commands `assign` and `unassign`.

1.4.1 *The* solve *command*

The `solve` command can be used to rearrange simple algebraic expressions to arrive at a new expression. For example, the solution of the equation

$$\frac{1}{x} = a + \frac{b}{x} \tag{1.1}$$

for x can be obtained using Maple as follows:

```
>   eq := 1/x=a+b/x;
```

$$eq := \frac{1}{x} = a + \frac{b}{x}$$

```
>   solve(eq,x);
```

$$-\frac{-1+b}{a}$$

Important

Note the different uses of `:=` and `=` here; `eq` is assigned the Maple version of Equation (1.1).

The command `solve(eq,x)` has solved the equation assigned to `eq` for the variable `x`; had we wanted to solve for `b` instead, we would have typed `solve(eq,b)`. In general, then, the syntax for solving one or more equations is `solve(S,X)` where `S` is either a single equation or a set of equations, and `X` is the required variable or set of variables. So suppose that we wish to solve the simultaneous equations

$$x + y = 2, \quad -x + 3y = 3.$$

This is done within Maple as follows:

```
>   restart:
>   eq := {x+y=2,-x+3*y=3}:
>   sol := solve(eq,{x,y});
```

$$sol := \{y = \frac{5}{4}, \, x = \frac{3}{4}\}$$

and a (rather clumsy) verification that this is correct can be obtained by evaluating `eq` at the solution point:

```
>   eval(eq,sol);
```

$$\{3 = 3, \, 2 = 2\}$$

We shall not always include the `restart:` line when we present code, but you should remember that whenever you are starting a piece of Maple input, it is advisable to include this line.

Note also that the curly brackets in the code here are essential and denote a set; as mentioned earlier, sets will be discussed in more detail in *Unit 2*.

Note that at this stage the variables x and y have not actually been assigned the values contained within the solution `sol`. This is confirmed by typing

```
>   x,y;
```

$$x, \, y$$

to see that x and y are returned unassigned. If we wished to assign x and y the values given in the solution, then we could do so by typing

```
>   x := 3/4:  y := 5/4:
>   x,y;
```

$$\frac{3}{4}, \frac{5}{4}$$

However, this is clearly long-winded and error-prone; a neater way is to use the `assign` command as follows:

```
> restart:  eq := {x+y=2,-x+3*y=3}:  u := {x,y}:
> sol := solve(eq,u);
```

$$sol := \{y = \frac{5}{4}, \ x = \frac{3}{4}\}$$

```
> assign(sol):
> x,y;
```

$$\frac{3}{4}, \frac{5}{4}$$

To understand what has happened here, type the following:

```
> restart:  assign(a=3);
> a;
```

$$3$$

In this simple example, `a` has been assigned the value 3 by using the command `assign(a=3)`, the effect of which is the same as if we had typed `a:=3` (except that even with a semicolon after it, `assign` does not output anything). To assign `b` the value 4, and then `c` the value `b`, type

```
> assign({b=4,c=b});
> b,c;
```

$$4, 4$$

The curly brackets around `b=4,c=b` are not necessary here, but have been included to be consistent – this is what happened with `sol` above, where `y` was assigned the value $\frac{5}{4}$ and `x` was assigned the value $\frac{3}{4}$.

In general, `assign(A=B)` can be thought of as having the same effect as `A := B`, with one big difference. Once `A` has been assigned a value, then `assign` as illustrated above cannot be used on `A` again; `A` has to be unassigned first.

To restore `A` to its unassigned status, use the **unassign** command:

```
> unassign('A');
```

Alternatively, a variable can be unassigned by re-assigning it to its name, that is, by typing `A := 'A':` (and again the single end of quote marks are essential).

Here the symbol ' is the single 'end of quote' mark, found on the same key as @ on a standard British keyboard. Its use is essential; type `?unassign` for further details.

The following exercises give you some practice in the use of `solve` and `assign`.

Exercise 1.17

Find the solutions of the following equations.

(a) $x^2 - x - 2025 = 0$

(b) $x^3 - 6x^2 - 19x + 24 = 0$

(c) $2x^4 - 11x^3 - 20x^2 + 113x + 60 = 0$

(Please be sure to enter the expressions in parts (b) and (c) carefully; Maple needs to be able to factorize the polynomials, and may not be able to do so if you make a mistake when typing them in.)

Exercise 1.18

Use the `solve` command to find the point of intersection, in the xy-plane, of the two lines

$$ax + by = A, \quad cx + dy = B.$$

Using `assign`, find an expression for the distance of the point of intersection from the origin, namely $\sqrt{x^2 + y^2}$.

1.4.2 The `fsolve` command

The `solve` command is used for finding *symbolic* solutions to equations. This may not always be possible, but nevertheless *numerical* solutions can usually be found with the very useful `fsolve` command. For example, suppose we wish to solve the equation

$$\sin x = x^3 - 5x^2 + 4,$$

and having plotted the graphs of $\sin x$ and $x^3 - 5x^2 + 4$, we know that there are three solutions, at approximately $x = -0.90$, 0.89 and 4.78. To find the negative solution more precisely, type

These approximate solutions were obtained in Exercise 1.15.

```
>   fsolve(sin(x)=x^3-5*x^2+4,x=-1..0);
                    -0.9004002089
```

The negative solution is $x = -0.900\,400\,208\,9$ to ten significant figures. Note that the range is specified using the `x=-1..0` construction; although this is optional, it is always advisable to assist Maple by giving it a range in which to search for solutions. The other two solutions of $\sin x = x^3 - 5x^2 + 4$ can be found by specifying different ranges:

The solution is given to ten significant figures because that is the current (i.e. the default) value of `Digits`.

```
>   fsolve(sin(x)=x^3-5*x^2+4,x=0..1);
                    0.8854364919
```

```
>   fsolve(sin(x)=x^3-5*x^2+4,x=4..5);
                    4.781398393
```

To test your understanding of `fsolve`, try the following exercises.

Exercise 1.19

Find real solutions of the equation

$$x \sin x = \tfrac{1}{2}$$

for x in the following ranges.

(a) $0 < x < 2$ (b) $2 < x < 4$ (c) $6 < x < 7$ (d) $8 < x < 10$

Verify graphically that there are no solutions in the range $4 < x < 6$.

Exercise 1.20

In Exercise 1.16, the five real solutions of $x \tan x = 1$ for $0 < x < 15$ were found approximately. Use the `fsolve` command to find each root correct to six decimal places.

Exercise 1.21

(a) Use `plot` to find the number and approximate value(s) of the real solution(s) of $x^3 + 3x^2 - 2x + 1 = 0$. Then use `fsolve` to find the solution(s) correct to six decimal places.

(b) Use `solve` to evaluate any remaining solutions of $x^3 + 3x^2 - 2x + 1 = 0$ that `fsolve` was unable to find.

(Note that in Exercise 1.21 one could have used `fsolve` with the option `complex` to find *all* the solutions of the given polynomial. Try the command `fsolve(x^3+3*x^2-2*x+1=0,x,complex)` or type `?fsolve,details` at the prompt for more information.)

You should now consolidate all you have learned so far by working through the next subsection.

1.4.3 An extended exercise – introducing hyperbolic functions

In the first four sections of this unit you have been introduced to some of the basic ideas within Maple. You have also seen how to evaluate expressions, plot simple line graphs and solve equations. In this subsection all this material is brought together in an extended exercise. The subject of this exercise is the family of *hyperbolic functions* which is made up of $\sinh x$, $\cosh x$ and $\tanh x$, and their respective reciprocals $\operatorname{cosech} x$, $\operatorname{sech} x$ and $\coth x$. You may have seen these functions before, but we assume that you have not.

Hyperbolic functions are introduced briefly in the course MS221.

Consider the exponential functions e^x and e^{-x} (where x may be complex). The *hyperbolic sine* function, written as $\sinh x$, is defined by

$$\sinh x = \tfrac{1}{2}(e^x - e^{-x}). \tag{1.2}$$

sinh is usually pronounced 'shine'.

Similarly, the *hyperbolic cosine* function, written as $\cosh x$, is defined by

$$\cosh x = \tfrac{1}{2}(e^x + e^{-x}). \tag{1.3}$$

cosh is pronounced as it is written, to rhyme with 'gosh'.

Since e^x and e^{-x} are defined for all complex x, so are $\sinh x$ and $\cosh x$.

These are built-in Maple functions, input as `sinh(x)` and `cosh(x)`.

Exercise 1.22

(a) Plot the graphs of $\sinh x$ and $\cosh x$ on the same figure for $-2 \le x \le 2$. Include a legend on your figure. Then spend a few moments analysing the graphs and note any features of interest. For example, are there any stationary points? What are the limiting values as $|x| \to \infty$? Do the functions $\sinh x$ and $\cosh x$ appear to be odd, even or neither? Are these observations consistent with the definitions (1.2) and (1.3) above?

The *hyperbolic tangent* $\tanh x$ is defined by

$$\tanh x = \frac{\sinh x}{\cosh x}, \qquad (1.4)$$

and input as `tanh(x)` in Maple.

tanh is usually pronounced 'tanch', i.e. 'tan' with a 'ch' for 'cheese' sound at the end.

(b) Given the definition of $\tanh x$, think about what form you expect the graph of $\tanh x$ to take. Then use Maple to plot the graph of $\tanh x$ for $-5 \le x \le 5$. Does it look as you expected it to?

(c) Evaluate the following to 15 significant figures:

$$\tanh 5, \quad \tanh 10, \quad \tanh 15, \quad \tanh 20.$$

(d) Find the real values of x for which $12 \cosh^2 x + 7 \sinh x = 24$.

The names of the hyperbolic functions suggest that they have certain properties in common with the trigonometric functions $\sin x$, $\cos x$, etc. This is indeed the case, and once you have completed this unit and learned a few more commands you will investigate this further.

1.5 Differentiation

Maple can easily find the derivatives of expressions of one or several variables. In this section we introduce the `diff` command, which is the basic differentiation operator for expressions, and illustrate its use in various situations including finding the coordinates of stationary points.

1.5.1 The `diff` command

The basic Maple command for differentiating expressions is `diff`, the use of which is best illustrated with some simple examples. The derivative of $\sin x$ with respect to x is found with the command

```
>  diff(sin(x),x);
```
$$\cos(x)$$

The general syntax for differentiating an expression `expr` with respect to a variable `var` is `diff(expr,var)`. Hence the derivative of $\cos y$ with respect to y is

```
>  diff(cos(y),y);
```
$$-\sin(y)$$

and the derivative of e^{x^2+4} with respect to x can be found with the commands

```
>  z := exp(x^2+4);
```
$$z := \mathrm{e}^{(x^2+4)}$$

```
>  d1 := diff(z,x);
```
$$d1 := 2\,x\,\mathrm{e}^{(x^2+4)}$$

The second derivative of e^{x^2+4} with respect to x is

```
>  diff(d1,x);
```

$$2\,\mathrm{e}^{(x^2+4)} + 4\,x^2\,\mathrm{e}^{(x^2+4)}$$

and this same result can also be found all in one go by typing

```
>  diff(z,x,x);
```

$$2\,\mathrm{e}^{(x^2+4)} + 4\,x^2\,\mathrm{e}^{(x^2+4)}$$

or the equivalent

```
>  diff(z,x$2);
```

$$2\,\mathrm{e}^{(x^2+4)} + 4\,x^2\,\mathrm{e}^{(x^2+4)}$$

where `x$2` is shorthand for `x,x` (try typing just `x$5;` at the prompt).

So the fifth derivative of $\sin(x^2)$ is achieved by typing
`diff(sin(x^2),x,x,x,x,x);` or with the command

```
>  diff(sin(x^2),x$5);
```

$$32\cos(x^2)\,x^5 + 160\sin(x^2)\,x^3 - 120\cos(x^2)\,x$$

Partial differentiation of expressions of more than one variable is performed by a straightforward extension of this procedure. For example, consider the expression $\sin x \cos y$; its second derivatives are found as follows. First define $z = \sin x \cos y$:

```
>  z := sin(x)*cos(y);
```

$$z := \sin(x)\cos(y)$$

Then the second derivative $\partial^2 z/\partial x^2$ is given by

```
>  diff(z,x$2);
```

$$-\sin(x)\cos(y)$$

whilst $\partial^2 z/\partial y^2$ can be found with

```
>  diff(z,y$2);
```

$$-\sin(x)\cos(y)$$

For the mixed second derivative $\partial^2 z/\partial x\partial y$, type

```
>  diff(z,x,y);
```

$$-\cos(x)\sin(y)$$

and to verify that the order of differentiation does not matter:

```
>  diff(z,y,x);
```

$$-\cos(x)\sin(y)$$

You may remember that the order of differentiation in a mixed derivative does not matter for sufficiently smooth functions.

If we wanted $\partial^5 z/\partial x^2 \partial y^3$, for example, then we would type

```
>  diff(z,x$2,y$3);
```

$$-\sin(x)\sin(y)$$

To illustrate an application of `diff`, define y to be the expression $x^2 + 4x + 7$, and find the coordinates of its stationary point as follows:

```
>  y := x^2+4*x+7;
```

$$y := x^2 + 4\,x + 7$$

Then the stationary point (that is, the point at which $dy/dx = 0$) can be found by typing

```
>   d1 := diff(y,x);
```

$$d1 := 2x + 4$$

```
>   solve(d1=0,x); eval(y,x=%);
```

$$\frac{-2}{3}$$

So the stationary point occurs at $(-2, 3)$.

The following exercises give further practice with the `diff` command and also revise `plot`, `solve` and `fsolve`. You should try to work through all the exercises; even though some questions appear similar, the methods required for their solutions differ.

Exercise 1.23

For each of the functions given below, use Maple to draw its graph over the range given. Then find the (x, y) coordinates of any stationary points.

(a) $y = 12x^5 - 15x^4 + 20x^3 - 330x^2 + 600x + 2, \quad -1 \le x \le 3$

(b) $y = \sin(x^{\cos x}), \quad 0 \le x \le 2\pi$

Exercise 1.24

Define the functions

$$y_1 = x^x, \quad y_2 = x^{(x^x)}, \quad y_3 = x^{(x^{(x^x)})},$$

and draw the graphs of y_1, y_2 and y_3 on the same plot for $0 \le x \le 1$. Use `solve` to show that y_1 has a minimum at $x = 1/e$, and use `fsolve` to show that y_3 has a minimum at $x = 0.2747$ (to four decimal places).

Exercise 1.25

Find the stationary points of the function

$$f(x, y) = 2x^2 - xy - 3y^2 - 3x + 7y - (x - 1)^3,$$

and classify the stationary point at $(1, 1)$.

[*Hint*: Remember that stationary points occur where $f_x = f_y = 0$. One method of classification is to use the $AC - B^2$ test. Evaluate $AC - B^2 = f_{xx}f_{yy} - (f_{xy})^2$ at the stationary point, which will be a saddle if $AC - B^2 < 0$, a local minimum if both $AC - B^2 > 0$ and $A > 0$, and a local maximum if both $AC - B^2 > 0$ and $A < 0$ (the test is inconclusive if $AC - B^2 = 0$).]

Here, f_x denotes $\partial f/\partial x$, etc. The $AC - B^2$ test is introduced in the course MST209; $A = f_{xx}(= \partial^2 f/\partial x^2)$, $B = f_{xy}$ and $C = f_{yy}$.

Exercise 1.26

If $z = \ln(x^2 + y^2)$, find the value of the constant a such that

$$\left(\frac{\partial z}{\partial x}\right)^2 + \left(\frac{\partial z}{\partial y}\right)^2 = ae^{-z}.$$

Sometimes it is necessary to find the derivative of an expression at a particular point, rather than as a function of the independent variable(s). This is achieved with a combination of the `eval` and `diff` commands. Thus the value of the first derivative of $\sin x$ at $x = 1.6$ can be found using the single command

```
> eval(diff(sin(x),x),x=1.6);
```
$$-0.02919952230$$

Recall that the Taylor series about $x = a$ for a function f of a single variable x is given by

$$f(a) + (x - a)f'(a) + \frac{1}{2!}(x - a)^2 f''(a) + \cdots$$

(where f has infinitely many continuous derivatives near $x = a$).

One method of finding the first four terms of the Taylor series of $y = \sin(e^x - 1)$ about $x = 0$ is to use the following three commands to compute the values of the first three derivatives of y at $x = 0$:

> Another method of finding Taylor series with Maple is to use the `series` command, details of which may be found by typing `?series` at the prompt. But that defeats the object here!

```
> y1 := eval(diff(sin(exp(x)-1),x),x=0);
```
$$y1 := 1$$

```
> y2 := eval(diff(sin(exp(x)-1),x$2),x=0);
```
$$y2 := 1$$

```
> y3 := eval(diff(sin(exp(x)-1),x$3),x=0);
```
$$y3 := 0$$

and, since $y(0) = 0$, the first few terms of the Taylor series are found using

```
> y := 0+y1*x+y2*x^2/2!+y3*x^3/3!;
```
$$y := x + \frac{1}{2}x^2$$

with the next term proportional to x^4 (or possibly a higher power of x).

> You may wish to check for yourself that the fourth derivative (evaluated at 0) is not zero.

Exercise 1.27

Use the `eval` and `diff` commands to find the first three non-zero coefficients of the Taylor series of $\sin(x + x^2)$ about $x = 0$ and about $x = \pi$. Then use Maple to find the first three non-zero terms of each series.

Exercise 1.28

Let

$$f = \ln\left(\frac{1 + \sin x}{\cos x}\right) - 2\sin x + x\cos x.$$

By evaluating the first five derivatives of f at $x = 0$, show that the first term in the Taylor series of f is $\frac{1}{15}x^5$.

1.5.2 Differentiation of arbitrary functions

So far only the derivatives of expressions with an explicit form (in terms of one or more independent variables) have been found. In this subsection we look at how to differentiate arbitrary functions, say $y(x)$ or $f(x, y)$, without first giving them an explicit form. For example, we might want to find dy/dx (as a function of y and x) given that $x^2 + y^2 = 3$. The problem is that typing

```
>   diff(y,x);
```

$$0$$

returns 0 (rather than the dy/dx that might have been expected) because Maple does not know that y depends on x; x and y are free independent variables and are treated as such.

This difficulty has been addressed in Maple by allowing the use of y(x); Maple interprets this syntax to mean that, for the purposes of differentiation and integration, y(x) depends on x. Thus we have

```
>   diff(y(x),x);
```

$$\frac{d}{dx}\, y(x)$$

```
>   diff(y(x),x$2);
```

$$\frac{d^2}{dx^2}\, y(x)$$

and indeed

```
>   diff(z(p),p);
```

$$\frac{d}{dp}\, z(p)$$

and even

```
>   diff(z(x,y),x);
```

$$\frac{\partial}{\partial x}\, z(x, y)$$

Maple interprets the syntax y(x) (or its equivalent in terms of other letters) in a manner consistent with the rules of calculus. With this notation, the **diff** command knows all the usual rules of differentiation, for instance the addition, multiplication and quotient rules:

```
>   diff(f(x)+g(x),x);
```

$$\left(\frac{d}{dx}\, f(x)\right) + \left(\frac{d}{dx}\, g(x)\right)$$

```
>   diff(f(x)*g(x),x);
```

$$\left(\frac{d}{dx}\, f(x)\right) g(x) + f(x) \left(\frac{d}{dx}\, g(x)\right)$$

```
>   diff(f(x)/g(x),x);
```

$$\frac{\frac{d}{dx}\, f(x)}{g(x)} - \frac{f(x)\left(\frac{d}{dx}\, g(x)\right)}{g(x)^2}$$

Maple also knows the rule for differentiating composite functions, such as $f(g(x))$, but gives the output in a notation not introduced until *Unit 2*.

To illustrate an application of this theory, suppose that we wish to find dy/dx (in terms of x and y) given that $x^2 + y^2 = 3$. To do this with Maple, first define the given explicit equation by typing

```
> eq1 := x^2+y(x)^2=3;
```
$$eq1 := x^2 + y(x)^2 = 3$$

and then differentiate this equation, assigning the output equation to a new name, eq2:

```
> eq2 := diff(eq1,x);
```
$$eq2 := 2\,x + 2\,y(x)\left(\frac{d}{dx}\,y(x)\right) = 0$$

Finally, rearrange this new equation to make dy/dx the subject:

```
> z := diff(y(x),x):   solve(eq2,z);
```
$$-\frac{x}{y(x)}$$

Hence $dy/dx = -x/y$.

Now try the following exercises.

Exercise 1.29

Find the derivatives of the following.

(a) $\sin(f(x))$ (b) $\sin\left(e^{f(x)}\right)$ (c) $\exp\left(\sqrt{1 + f(x)g(x)}\right)$

Exercise 1.30

Use the **solve** command to find dy/dx (in terms of y and x) given that

$$x^2 + y^2 + 3xy = 0.$$

In solving Exercise 1.30, you had to type y(x) several times, when it might have been tempting to type just y. It is actually possible to persuade Maple to allow you to do this. If you type alias(y=y(x)); then Maple will regard every subsequent occurrence of y as if it were y(x).

More details about the **alias** command may be found by typing ?alias at the prompt.

Exercise 1.31

Repeat Exercise 1.30 using the **alias** command.

Using **alias** in this way is potentially dangerous; it is important to remember that the alias is in place. However, as long as you do remember this, the use of **alias** can save typing and make the Maple output look more natural.

Exercise 1.32

Prove that $f(x - ct)$ satisfies the partial differential equation

$$c^2\frac{\partial^2 f}{\partial x^2} = \frac{\partial^2 f}{\partial t^2}.$$

That concludes the work on the differentiation operator **diff**. In the next section we stay with calculus and look at integration.

31

1.6 Integration

Maple can evaluate many integrals symbolically, and it can also provide numerical estimates of most definite integrals which cannot be evaluated symbolically as well as those that can. In this section the integration command `int` is introduced, along with the related `Int`. We also introduce the commands `assume` and `about`.

1.6.1 The `int` and `Int` commands

The `int` command is best explained with a simple example. To integrate the expression xe^{ax^2} with respect to x, type

```
>   int(x*exp(a*x^2),x);
```

$$\frac{1}{2}\frac{e^{(a x^2)}}{a}$$

so the syntax for integrating an expression `expr` with respect to a variable `var` is `int(expr,var)`. Note that Maple does not include an arbitrary constant of integration, which means that the same function may be integrated to a different expression depending on how it is expressed in Maple; for example, consider

```
>   int((1+x)^2,x);
```

$$\frac{(1+x)^3}{3}$$

```
>   int(1+2*x+x^2,x);
```

$$x + x^2 + \frac{1}{3}x^3$$

Now $\dfrac{(1+x)^3}{3} = \dfrac{1}{3} + x + x^2 + \dfrac{x^3}{3}$, so (as we would hope!) the two expressions differ only by a constant.

The limits on definite integrals are input using the `x=a..b` construction which you have seen before, in `plot` and `fsolve`, for example. So to evaluate the definite integral

$$\int_0^1 xe^{5x^2}\,dx,$$

type

```
>   int(x*exp(5*x^2),x=0..1);
```

$$-\frac{1}{10} + \frac{1}{10}e^5$$

and to evaluate

$$\int_0^u \frac{dx}{\sqrt{u-x}},$$

type

```
>   int(1/sqrt(u-x),x=0..u);
```

$$2\sqrt{u}$$

(In the `x=a..b` construction, `a` and `b` can be variables, numbers or expressions.)

If Maple cannot evaluate an integral, then it simply returns it:

```
>  int(x^x,x);
```

$$\int x^x \, dx$$

At this point you really need to experiment with some integrals, just to see what can be done.

Exercise 1.33

Use Maple to integrate the following expressions with respect to x.

(a) $\sqrt{e^x - 1}$

(b) $x^2(ax + b)^{5/2}$, where a and b are constants

(c) $\sinh 6x \sinh^4 x$

(d) $(\cosh x)^{-6}$

(e) $\sin(\ln x)$

Exercise 1.34

Use Maple to evaluate the following integrals symbolically.

(a) $\displaystyle\int_{1/2}^{1} \frac{1}{1 + x^3} \, dx$

(b) $\displaystyle\int_{0}^{1} x^2 \arctan x \, dx$

(c) $\displaystyle\int_{0}^{\infty} \frac{1}{(1 + x)(1 + x^2)} \, dx$

From the exercises you have just done you will observe that Maple's ability to do rather nasty integrals is impressive, and that it easily evaluates all of the integrals found in elementary texts as well as most of the indefinite integrals given in standard tables.

Nevertheless one can encounter cases where it is desirable to have Maple return an integral unevaluated rather than let it attempt to evaluate it symbolically. This is done using the **inert form** of the integration command, Int. By inert, we mean that the integral is not evaluated, it is merely returned as an integral. For example, type

```
>  Int(x^3/(1+x^2),x);
```

$$\int \frac{x^3}{1 + x^2} \, dx$$

Note that Maple just returns the integral unevaluated, whereas had the active form int been used, the integral would have been symbolically evaluated. An important use of Int is in numerical integration, as explained in the next subsection.

1.6.2   Numerical evaluation of definite integrals

If Maple fails to find a symbolic answer to a definite integral, or if you know that the integral in question cannot be evaluated in terms of known functions, then it may be necessary to evaluate it numerically. Maple does this by using the `evalf` function as illustrated in the following example, where we try to evaluate

$$\int_0^1 e^{-t} \arcsin t \, dt.$$

First, try to find a symbolic result by typing

```
>   z := int(exp(-t)*arcsin(t),t=0..1);
```

$$z := \int_0^1 e^{(-t)} \arcsin(t) \, dt$$

Maple just returns the integral, which means that it does not know any symbolic answer. So, instead, find the numerical value of the integral simply by evaluating `z` as a floating point number:

```
>   evalf(z);
```

$$0.2952205678$$

Alternatively, to do the whole thing in one go, and to four significant figures (say), type

```
>   evalf[4](Int(exp(-t)*arcsin(t),t=0..1));
```

$$0.2952$$

Note that the use of `Int` here, as opposed to `int`, saves Maple the time-consuming and futile attempt at evaluating the integral symbolically before giving the numerical solution.

Exercise 1.35

Use Maple to evaluate the following integrals numerically to six significant figures.

(a) $\displaystyle\int_0^1 e^{x^3} dx$ (b) $\displaystyle\int_0^{10} \frac{1}{\sqrt{1+x^4}} \, dx$ (c) $\displaystyle\int_0^5 \sin(e^{x/2}) \, dx$

The integration routine provided with Maple seems able to estimate accurately the values of most definite integrals whose integrands are well-behaved. It does, however, fail in extreme cases, such as when the integrand is very oscillatory. You will not come across such cases in this unit.

For example, Maple has trouble estimating

$$\int_0^\pi \frac{\cos(at^2)}{1+t^4} \, dt \text{ accurately, for}$$

large values of a.

1.6.3   Definite integration and the `assume` facility

Many definite integrals contain parameters which must lie within specific ranges for the integral to exist. For example, suppose that we wish to evaluate the integral

$$\int_0^\infty e^{-xy} \, dx$$

and type

```
>   int(exp(-x*y),x=0..infinity);
```

$$\lim_{x \to \infty} -\frac{e^{(-xy)} - 1}{y}$$

Maple, quite rightly, cannot evaluate this integral because it does not exist unless $y > 0$. This problem can be overcome by using the **assume** command as follows:

```
> assume(y>0);
> int(exp(-x*y),x=0..infinity);
```

$$\frac{1}{y\sim}$$

The expected result is now obtained; note the **tilde** after the variable y, which indicates that it has assumed properties. (The tilde can be turned off, but we do not recommend that you do this as it is useful to be reminded that assumptions are in place. Type **?showassumed** and scroll down to the 'showassumed' paragraph if you want to learn more about not displaying assumed variables.)

Another frequently recurring example is in the computation of the coefficients of Fourier series. Suppose we need the nth Fourier cosine coefficient of the function x^2 on the interval $(-\pi, \pi)$. This is given by

Fourier series are introduced in the course MST209.

$$\frac{1}{\pi} \int_{-\pi}^{\pi} x^2 \cos(nx)\, dx$$

You will explore this further in Exercise 1.57.

and is evaluated in Maple as follows:

```
> z := int(x^2/Pi*cos(n*x),x=-Pi..Pi);
```

$$z := \frac{2\left(-2\sin(n\pi) + n^2 \sin(n\pi)\pi^2 + 2n\cos(n\pi)\pi\right)}{\pi n^3}$$

In complicated expressions like this, your output may look slightly different from ours (for example, πn instead of $n\pi$).

This is clearly not as neat as it could be, because $\sin(n\pi) = 0$ for any integer n. Maple has not set these terms to zero because it does not know that n is an integer. This can be achieved by typing

```
> assume(n::integer);
> z;
```

$$\frac{4(-1)^{n\sim}}{n\sim^2}$$

The :: here is two colons, one after the other. It can be thought of as meaning 'is of type'.

Not only has Maple now set $\sin(n\pi) = 0$, but it has also set $\cos(n\pi) = (-1)^n$.

If in the above example we had wished to assume n to be a *positive* integer then this could have been achieved by typing **assume(n>0,n::integer)**. In other words, more than one assumed property can be placed inside the round brackets, each separated by a comma. (Alternatively, in this instance, **assume(n::posint)** is also possible; you will encounter this later in the course – see Table 3.1 on page 136.)

A useful command, related to **assume**, is **about** which tells us whether a variable has any properties associated with it, and if so what properties they are. Thus if a is required to be real and positive, type

```
> assume(a>0);
```

and if we subsequently forget what properties are associated with the variables a and b, then we can type

```
> about(a,b);
Originally a, renamed a~:
  is assumed to be:  RealRange(Open(0),infinity)
b:
  nothing known about this object
```

To clear the properties from the variable a, type a := 'a', as introduced in Subsection 1.4.1 (remember that ' is the 'end of quote' mark):

> a := 'a';

$$a := a$$

> about(a);

```
a:
  nothing known about this object
```

Using **unassign** here would *not* clear the properties from a.

One final point worth making about **assume** is that, when it is applied to a variable, any previous assumptions made about that variable are forgotten. This is illustrated in the following example:

> assume(a>1);

> assume(a::integer,a>-4);

> about(a);

```
Originally a, renamed a~:
  is assumed to be:  AndProp(integer,RealRange(-3,infinity))
```

The assume(a::integer,a>-4) command has overwritten assume(a>1), which is now forgotten.

If you want to know any more about **assume** or **about** then type ?assume or ?about at the prompt. The following exercise completes this section on integration.

Exercise 1.36

Evaluate the following integrals.

(a) $\displaystyle\int_1^n \frac{1}{x}\,dx$ (b) $\displaystyle\int_{-\infty}^{\infty} e^{ax^2}\,dx,$ for $a < 0$

1.7   *The* sum *and* add *commands*

In this section the similar, but different, **sum** and **add** commands are introduced; these are used to evaluate sums. First, we illustrate both commands with the evaluation of the sum S defined by

$$S = \sum_{r=1}^{10} r = 1 + 2 + 3 + \cdots + 9 + 10 = \tfrac{1}{2} \times 10 \times 11 = 55.$$

To find this using the **sum** command, type

> S := sum(r,r=1..10);

$$S := 55$$

The syntax for the sum command is sum(expr,r=a..b), where the first argument expr is the expression to be summed and the second argument r=a..b shows the variable r running from a to b (inclusive, incremented by 1). You have already seen the r=a..b construction, for example in the plot and int commands, although when used with sum the limits a and b must be integers and the increment is always 1.

The add command has a similar structure to sum and produces (thankfully) the same result:

```
>  A := add(r,r=1..10);
```
$$A := 55$$

If the limits in sum and add are not integers, a result is still produced (different in the two cases). You may wish to try the limits r=1.5..3.7 and try to work out what Maple is doing.

Let us now consider a slightly more general case, where the upper limit is replaced by a free variable, n. Then

$$S = \sum_{r=1}^{n} r = 1 + 2 + 3 + \cdots + (n-1) + n = \tfrac{1}{2}n(n+1). \tag{1.5}$$

Here you will see that the two commands behave quite differently. The sum command yields the standard result, albeit in a slightly different form:

```
>  S := sum(r,r=1..n);
```
$$S := \frac{(n+1)^2}{2} - \frac{n}{2} - \frac{1}{2}$$

You will see in the next section how to direct Maple to output the result in the factorized form given in Equation (1.5).

The add command, however, fails:

```
>  A := add(r,r=1..n);
```
```
Error, unable to execute add
```

The reason for this behaviour is that the sum command is for symbolic summation. It is used to compute a closed form for an indefinite or definite sum. The add command is used to add up an explicit sequence of numerical values and so will not work for indefinite sums. Whilst sum can often be used to compute definite sums (as seen above), it is strongly recommended that add be used if a definite sum is needed. Indeed, if asked to evaluate a simple finite sum using sum, Maple itself will usually use add to do so! (There are other differences between sum and add, but they are not of concern here. If you are interested in learning more, type ?sum and scroll down to 'Comparison of Sum and Add'.)

A definite sum is one with an explicit numerical range, and an indefinite sum is one that does not have an explicit numerical range.

Infinite sums are evaluated using sum and the Maple symbol for ∞; for example, the infinite geometric series is found by typing

```
>  assume(abs(a)<1):
>  sum(a^k,k=1..infinity);
```
$$-\frac{a\sim}{a\sim - 1}$$

(You should be aware that in this case Maple gives the same output regardless of whether $|a|$ is assumed less than 1 or not; this is an error in Maple as the sum to infinity is valid only if $|a| < 1$.)

With infinite sums, add will not work; try typing add(a^k,k=1..infinity); – you should obtain an error message.

Now try the following exercises (overleaf).

Exercise 1.37

Use Maple to evaluate the following.

(a) $\displaystyle\sum_{r=1}^{10}(r^2+3r-2)$ (b) $\displaystyle\sum_{i=1}^{50}2^i$ (c) $\displaystyle\sum_{k=0}^{n-1}a^k$ (d) $\displaystyle\sum_{k=0}^{\infty}\frac{1}{k!}$

Exercise 1.38

Use Maple to verify the following.

(a) $\displaystyle\sum_{n=1}^{\infty}\frac{2}{(n+1)(n+2)}=1$ (b) $\displaystyle\sum_{k=0}^{\infty}\frac{k^2+k-1}{(k+2)!}=0$ (c) $\displaystyle\sum_{k=1}^{\infty}\frac{1}{k^2}=\frac{\pi^2}{6}$

The `evalf` command can be used with `add` to evaluate finite sums numerically. For example, the sum

$$\sum_{k=1}^{100}\frac{1}{k+k^{3/2}}$$

can be evaluated with

```
>  add(evalf(1/(k+k^(3/2))),k=1..100);
                  1.493842552
```

(Using `add(evalf(...))` in this way, rather than `evalf(add(...))`, requires less memory for sums with many terms, though it risks decreasing numerical accuracy, particularly if positive and negative quantities are involved.)

Now try the following exercises.

Exercise 1.39

Find the numerical values of the following sums.

(a) $\displaystyle\sum_{k=1}^{10}e^{-\sqrt{k}}$ (b) $\displaystyle\sum_{k=1}^{100}\frac{1}{\sqrt{k}}$

Exercise 1.40

By embedding one `add` command within another, evaluate the following double sums to 10 significant figures.

(a) $\displaystyle\sum_{m=0}^{10}\sum_{n=0}^{10}(2n+1)e^m$ (b) $\displaystyle\sum_{n=-20}^{20}\sum_{m=-10}^{10}\frac{1}{n^2+\ln(22+m)}$

That almost concludes this section on the `sum` and `add` commands. However, you may be wondering if there are commands similar to `sum` and `add`, but for evaluating products. Maple does indeed have `product` and `mul`, and these are explored in the final exercise of this section.

Exercise 1.41

Assuming that `product` and `mul` behave in the same way as `sum` and `add`, respectively, use Maple to evaluate the following.

Remember that you can obtain more information by typing ?product and ?mul at the prompt.

(a) $\displaystyle\prod_{r=1}^{10} r^2$ (b) $\displaystyle\prod_{r=1}^{5}(1 - q^r)$

Note: $\prod$ denotes a product in the same way as $\sum$ denotes a sum. For example,

$$\prod_{r=1}^{5}(r + 2) = 3 \times 4 \times 5 \times 6 \times 7 = 2520.$$

1.8   Manipulation

The main strength of Maple lies in its ability to manipulate mathematical expressions symbolically. So far in this unit you have seen how to differentiate and integrate expressions, how to perform summations and how to solve equations. This section introduces the four main commands `expand`, `factor`, `simplify` and `combine` that enable mathematical expressions to be rearranged. Some of the examples given here are necessarily rather simple, so that you can gain familiarity with these new commands, but we will also revisit some of the earlier material in this unit and see how the Maple output may be improved.

The symbolic manipulation of expressions rarely creates results in the form needed by the user. This is because there are many ways of writing any expression. For example, $\cos(2x)$ might sometimes be better written as $2\cos^2 x - 1$, or as $1 - 2\sin^2 x$, or as $\cos^2 x - \sin^2 x$. Sometimes $x^3 - y^3$ is better as $(x - y)(x^2 + xy + y^2)$, but sometimes it is better left just as it is. Algebraic systems such as Maple have to apply fixed rules (for example, Maple almost always replaces $\sin^2 x$ with $1 - \cos^2 x$), and these rules may conflict with the requirements of the user. As you become more adept at using Maple, you need to know how to persuade it to change the form of an expression into that which you find the most useful. This is more of an art than a science, but a minimum requirement is a working knowledge of the commands and techniques introduced here. Usually some ingenuity is also needed, and plenty of experience, so the more you practise these important techniques, the better.

1.8.1   expand

The `expand` command does exactly what its name implies. In this subsection we will look at how it works when applied to polynomials and to trigonometric, exponential and logarithmic functions. The `coeff` command is also introduced.

The effect of **expand** on polynomial expressions is fairly obvious; for example, the expansion of $(x+1)(x+z)^2$ is found by typing

```
>   expand((x+1)*(x+z)^2);
```
$$x^3 + 2x^2z + xz^2 + x^2 + 2xz + z^2$$

If all that is required is a single coefficient, however, then the **coeff** command is perhaps more useful. For example, the coefficient of x^3 in the expansion of $(x+1)(x+2)^3(x-2)^4$ can be found as follows:

```
>   p := (x+1)*(x+2)^3*(x-2)^4;
```
$$p := (x+1)(x+2)^3(x-2)^4$$

```
>   coeff(p,x^3);
```
$$-48$$

```
>   expand(p);
```
$$128 + 64x - 48x^3 + x^8 - x^7 - 14x^6 + 12x^5 - 160x^2 + 72x^4$$

(The final command is just to verify the first result.)

When applied to trigonometric functions, **expand** uses the sum rules to remove multiple angles, replacing them with powers:

An example of a sum rule is $\cos(x \pm y) = \cos x \cos y \mp \sin x \sin y$.

```
>   expand(cos(2*x));
```
$$2\cos(x)^2 - 1$$

Other examples of the **expand** command are

```
>   expand(sin(x+y));
```
$$\sin(x)\cos(y) + \cos(x)\sin(y)$$

```
>   expand(cos(5*x));
```
$$16\cos(x)^5 - 20\cos(x)^3 + 5\cos(x)$$

```
>   coeff(expand(cos(5*x)),cos(x)^5);
```
$$16$$

The final command line here illustrates how **coeff** may be used in trigonometric expansions. Note that it is necessary to expand $\cos(5x)$ first; try typing `coeff(cos(5*x),cos(x)^5);` to see why.

If in the last expansion above we wished instead to obtain the partial expansion $\cos(5x) = \cos(4x)\cos x - \sin(4x)\sin x$, then we would need a little trickery:

```
>   c := expand(cos(x+y));
```
$$c := \cos(x)\cos(y) - \sin(x)\sin(y)$$

```
>   eval(c,y=4*x);
```
$$\cos(x)\cos(4x) - \sin(x)\sin(4x)$$

To test your understanding of **expand**, try the following exercise.

Exercise 1.42

Use Maple to show the following.

(a) $\sin(3x) = 4\sin x \cos^2 x - \sin x$

(b) $\sin(3x) = \sin x \cos(2x) + \sin(2x) \cos x$

Maple also expands hyperbolic expressions in much the same manner; for example, try `expand(sinh(3*x));`.

The operation of `expand` on the exponential function is straightforward:

> `expand(exp(x+y));`
$$\mathbf{e}^x \, \mathbf{e}^y$$

> `expand(exp(z*(x+y)));`
$$\mathbf{e}^{(z\,x)} \, \mathbf{e}^{(z\,y)}$$

Some care is needed with the logarithm function ln, but if its arguments are assumed to be real and positive, then `expand` produces the standard logarithmic rules:

> `assume(x>0,y>0);`

> `expand(ln(x*y));`
$$\ln(x\sim) + \ln(y\sim)$$

> `expand(ln(x/y));`
$$\ln(x\sim) - \ln(y\sim)$$

> `expand(ln(x^y));`
$$y\sim \ln(x\sim)$$

(Without the assumptions made on x and y at the start, Maple would generally assume them to be complex and would not perform these manipulations. This causes difficulties with ln which we do not need to go into here.)

That completes this brief introduction to `expand`. Now we move on to look at `factor`.

1.8.2 `factor`

For polynomials, the `factor` command is in some ways the opposite of `expand`, and once more it does what one would expect from its name:

> `factor(x^5-x^4-7*x^3+x^2+6*x);`
$$x\,(x-1)\,(x-3)\,(x+2)\,(x+1)$$

> `f := x^4+4*x^3*y-7*x^2*y^2-22*x*y^3+24*y^4;`
$$f := x^4 + 4\,x^3\,y - 7\,x^2\,y^2 - 22\,x\,y^3 + 24\,y^4$$

> `factor(f);`
$$(-y+x)\,(-2\,y+x)\,(3\,y+x)\,(4\,y+x)$$

The `factor` command can also be used on rational expressions:

> `factor((x^3-y^3)/(x^4-y^4));`
$$\frac{x^2 + x\,y + y^2}{(y+x)\,(x^2+y^2)}$$

(Maple has implicitly assumed here that $x \neq y$ and has cancelled the common factor $x - y$ from the numerator and denominator.)

The following exercises, which conclude this subsection, give more practice on the use of `factor` and also introduce the command `ifactor` which is used for the prime factorization of an integer.

Exercise 1.43

Expand the function $f = (x + y + 1)^2$, and then factorize $f - 1$.

Exercise 1.44

Use `expand` and `factor` to find values of the constants p, q, r, s, t and u such that

$$\frac{\sinh 5x}{\sinh x} = (p \cosh^2 x + q \cosh x + r)(s \cosh^2 x + t \cosh x + u).$$

Exercise 1.45

Use Maple to show that

$$\sum_{r=1}^{n} r = \tfrac{1}{2} n(n + 1),$$

$$\sum_{r=1}^{n} r^2 = \tfrac{1}{6} n(n + 1)(2n + 1),$$

$$\sum_{r=1}^{n} r^3 = \tfrac{1}{4} n^2 (n + 1)^2.$$

The first of these three results was given in Equation (1.5) on page 37. It was stated there that later on you would learn how to use Maple to obtain it.

Exercise 1.46

The command `ifactor` acts on an integer and factorizes it into its constituent primes. For example:

```
>   ifactor(12);
```

$$(2)^2 \ (3)$$

Use Maple to find the prime factorization of 10! and of 10! + 1. Show that 11! + 1 is a prime number.

1.8.3 simplify

The command `simplify` is the most general of Maple's simplification commands and is usually the one to be tried first. However, it is not always the most appropriate command and its results can be unpredictable, so you are advised to keep a copy of the original expression. Normally, however, it does produce a simpler result, and it does so by applying the following rules (which you need not remember):

$$
\begin{aligned}
\sin^2 x &\;\to\; 1 - \cos^2 x, \\
\sinh^2 x &\;\to\; \cosh^2 x - 1, \\
\exp(x)\exp(y) &\;\to\; \exp(x + y), \\
u^x u^y &\;\to\; u^{x+y}, \\
u^{a/b} u^{c/d} &\;\to\; u^{(ad+bc)/bd}, \\
\ln(xy) &\;\to\; \ln x + \ln y, \text{ but only if } x > 0 \text{ or } y > 0, \\
(xy)^z &\;\to\; x^z y^z, \text{ but only if } x > 0 \text{ or } y > 0.
\end{aligned}
$$

Positive integer powers of trigonometric and hyperbolic functions are simplified by these rules as far as possible, as illustrated by the following request to simplify $\sinh^4 x - \cosh^4 x$:

```
>  simplify((sinh(x))^4-(cosh(x))^4);
```
$$1 - 2\cosh(x)^2$$

One of the most useful applications of `simplify` is when trying to verify that `expr1 = expr2`, where `expr1` and `expr2` are both expressions. This is achieved with the command `simplify(expr1-expr2)`, as illustrated by the following verification that

$$\frac{1 + \tanh^2 x}{1 - \tanh^2 x} = \cosh 2x.$$

```
>  expr1 := (1+tanh(x)^2)/(1-tanh(x)^2);
```
$$expr1 := \frac{1 + \tanh(x)^2}{1 - \tanh(x)^2}$$

```
>  expr2 := cosh(2*x);
```
$$expr2 := \cosh(2\,x)$$

```
>  simplify(expr1-expr2);
```
$$0$$

The following exercises give you some practice in applying the `simplify` command. You may also need to use some earlier commands such as `solve` and `diff`.

Exercise 1.47

Use the `simplify` command to show the following.

(a) If $y = \dfrac{1 + \sin x}{1 + \cos x}$, then $\dfrac{dy}{dx} = \dfrac{\cos x + \sin x + 1}{1 + 2\cos x + \cos^2 x}$.

(b) If $y = \ln\sqrt{\dfrac{1 + x}{1 - x}}$, then $\dfrac{dy}{dx} = -\dfrac{1}{-1 + x^2}$.

(c) If $y = \ln\left(\dfrac{(1 + x)^{1/2}}{(1 - x)^{1/3}}\right)$, then $\dfrac{dy}{dx} = \dfrac{-5 + x}{6(-1 + x^2)}$.

Exercise 1.48

Show that the function $f = 1/r$, where $r^2 = (x - a)^2 + (y - b)^2 + (z - c)^2$ and a, b and c are constants, is a solution of Laplace's equation

$$\frac{\partial^2 f}{\partial x^2} + \frac{\partial^2 f}{\partial y^2} + \frac{\partial^2 f}{\partial z^2} = 0.$$

Sometimes the `simplify` command does not do what one might expect, as the following example illustrates.

At the prompt, type the following:

```
>  simplify(sqrt(x)*sqrt(1/x));
```
$$\sqrt{x}\,\sqrt{\frac{1}{x}}$$

Observe that the output is not 1, as might have been expected (although a result of 1 can be obtained if `assume` is used to make $x > 0$ first). In situations such as this, the 'natural' simplification can sometimes be forced by using `simplify` with an optional argument, `symbolic`. Typing

```
>  simplify(sqrt(x)*sqrt(1/x),symbolic);
```
$$1$$

produces the 'natural' result automatically. The reason is that `symbolic` allows Maple to manipulate $\sqrt{x}$ and $\sqrt{1/x}$ in an algebraically straightforward way without taking account of the possibility that x may not be positive.

Another example is

```
>  simplify(arctan(tan(x)));
```
$$\arctan(\tan(x))$$

which has no unique simplified form because $\tan(n\pi + x) = \tan x$ for any integer n. However, frequently it is not necessary to worry about this complication and it is satisfactory to accept the result x.

Exercise 1.49

Use `simplify` with `symbolic` to obtain a simplification of $\arctan(\tan(x))$.

1.8.4 combine

You saw in Subsection 1.8.2 that `factor` is in some ways the opposite of `expand` for polynomials. In other circumstances it is more the case that `combine` performs this opposite operation. The following examples show how it can be used.

> Further examples can be found by typing `?combine` at the prompt.

```
>  combine(exp(x)*exp(y));
```
$$\mathbf{e}^{(x+y)}$$

```
>  combine(x^u*x^v);
```
$$x^{(u+v)}$$

```
>  combine(ln(x)+ln(y));
```
$$\ln(x) + \ln(y)$$

```
>  assume(x>0,y>0):  combine(ln(x)+ln(y));
```
$$\ln(x{\sim} y{\sim})$$

```
>  x := 'x':  y := 'y':
```

> Recall that we also used $x > 0$ and $y > 0$ when using `expand` on $\ln(xy)$.
>
> Once you have seen the result of `combine(ln(x)+ln(y));`, you no longer need the assumptions $x > 0$ and $y > 0$, so you should clear them as explained on page 36.

The following example shows how expressions involving rational powers can be simplified:

```
>  assume(p::integer, q::integer, r::integer, s::integer):
>  combine(x^(p/q)*x^(r/s));
```
$$x^{\left(\frac{p{\sim}}{q{\sim}}+\frac{r{\sim}}{s{\sim}}\right)}$$

One of the useful applications of `combine` is in trigonometric simplification, because it transforms powers of sin and cos into their multiple-angle expansions. For example:

```
>  combine(sin(x)^2);
```
$$\frac{1}{2} - \frac{1}{2}\cos(2x)$$

```
>  combine(cos(x)^6);
```

$$\frac{1}{32}\cos(6\,x) + \frac{3}{16}\cos(4\,x) + \frac{15}{32}\cos(2\,x) + \frac{5}{16}$$

```
>  combine(sin(x)^2*(1-cos(x)^5));
```

$$\frac{1}{2} + \frac{3}{64}\cos(5\,x) + \frac{1}{64}\cos(3\,x) - \frac{5}{64}\cos(x) - \frac{1}{2}\cos(2\,x) + \frac{1}{64}\cos(7\,x)$$

Exercise 1.50

Use Maple to find multiple-angle forms for $\sin^7 x$, $\cos^3 x$ and $\cosh^4 x$.

1.8.5 `simplify` *with side relations*

In this subsection we introduce a very helpful feature of the `simplify` command, the *side relation*. This will be illustrated within the context of exploring some features of the series

$$S_p = \sum_{k=1}^{n} k^p,$$

where p is a positive integer. It transpires that there are many relationships between the S_p for various values of p, for instance $S_3 = S_1^2$, as you may have spotted when working through Exercise 1.45. Before we explain what side relations are, and look at their use in establishing some relationships between the S_p, you should complete the following exercise.

Exercise 1.51

Define S_p by typing

```
>  Sp := sum(k^p,k=1..n);
```

$$Sp := \sum_{k=1}^{n} k^p$$

so that S_1 (in factorized form) can be found easily with the command

```
>  S1 := factor(eval(Sp,p=1));
```

$$S1 := \frac{n\,(n+1)}{2}$$

(a) In a similar manner, show that

$$S_2 = \tfrac{1}{6}n(n+1)(2n+1),$$
$$S_3 = \tfrac{1}{4}n^2(n+1)^2,$$
$$S_4 = \tfrac{1}{30}n(n+1)(2n+1)(3n^2+3n-1),$$
$$S_5 = \tfrac{1}{12}n^2(n+1)^2(2n^2+2n-1),$$
$$S_7 = \tfrac{1}{24}n^2(n+1)^2(3n^4+6n^3-n^2-4n+2).$$

(b) Use the `simplify` command to verify that

$$S_3 = S_1^2, \quad S_5 = \tfrac{1}{3}S_1^2(4S_1-1) \quad \text{and} \quad S_7 = \tfrac{1}{3}S_1^2(6S_1^2-4S_1+1).$$

The results of Exercise 1.51(b) show how S_3, S_5 and S_7 are related to S_1. Now you will see how Maple can be used to *derive* these, and more, relationships rather than just verify them.

This is achieved using the idea of a constraint, or *side relation* (see below), which is used with the `simplify` command. In order to derive the relationship $S_3 = S_1^2$, you should first set things up again from the beginning (in case anything was altered whilst working through the above exercise):

```
>  restart:    Sp := sum(k^p,k=1..n):
>  S1 := factor(eval(Sp,p=1)):   S3 := factor(eval(Sp,p=3)):
```

Then the required result can be found by typing

```
>  S3 := simplify(S3,{S1=a1});
```

$$S3 := a1^2$$

where this command may be read as 'simplify S3 subject to the constraint that S1=a1 and assign the result to S3' (so that the value of S3 is overwritten). The **side relation** S1=a1 is the second argument in the `simplify` command, and must be enclosed in curly brackets. The output $S3 := a1^2$ gives the result $S_3 = S_1^2$ because the side relation has a1 equal to S1.

So, to derive the relation between S_5 and S_1, type

```
>  S5 := factor(eval(Sp,p=5));
```

$$S5 := \frac{n^2\,(2\,n^2 + 2\,n - 1)\,(n+1)^2}{12}$$

```
>  S5 := simplify(S5,{S1=a1});
```

$$S5 := \frac{4}{3}\,a1^3 - \frac{1}{3}\,a1^2$$

and then factorize the output:

```
>  factor(S5);
```

$$\frac{a1^2\,(4\,a1 - 1)}{3}$$

To test your understanding of side relations, try the following exercise.

Exercise 1.52

Use Maple to derive the relation for S_7 in terms of S_1 given in Exercise 1.51(b), and to derive the following.

(a) $S_9 = \frac{1}{5}S_1^2(2S_1 - 1)(8S_1^2 - 6S_1 + 3)$

(b) $S_{11} = \frac{1}{3}S_1^2(16S_1^4 - 32S_1^3 + 34S_1^2 - 20S_1 + 5)$

The expressions for S_2 and S_4 found in Exercise 1.51(a) are respectively cubic and quintic in n, so they cannot be expressed as polynomials in S_1 (which is quadratic in n). But it seems possible that S_4 may be expressible in terms of S_1 *and* S_2. This can be tested by extending the side relation to include the equation S2=a2:

```
>  Sp := sum(k^p,k=1..n):   S1 := factor(eval(Sp,p=1)):
>  S2 := factor(eval(Sp,p=2)):   S4 := factor(eval(Sp,p=4)):
>  S4 := factor(simplify(S4,{S1=a1,S2=a2}));
```

$$S4 := \frac{a2\,(6\,a1 - 1)}{5}$$

The result $S_4 = S_2(6S_1 - 1)/5$ is obtained.

Exercise 1.53

Show the following.

(a) $S_6 = \frac{1}{7}S_2(12S_1^2 - 6S_1 + 1)$

(b) $S_8 = \frac{1}{5}S_2(15S_2^2 - 15S_1^2 + 6S_1 - 1)$ (Maple 10)   Maple 11 gives $S_8 = \frac{1}{15}S_2(40S_1^3 - 40S_1^2 + 18S_1 - 3)$

To end this subsection on side relations, try the following exercise.

[*Hint*: Side relations are equations, so it is permissible to assign a set of side relations a name (in the same way that the pair x+y=2,-x+3*y=3 in Subsection 1.4.1 was assigned the name eq).]

Exercise 1.54

If $a + b + c = u$, $a^2 + b^2 + c^2 = v$ and $a^3 + b^3 + c^3 = w$, show the following.

(a) $a^4 + b^4 + c^4 = \frac{1}{6}u^4 + \frac{1}{2}v^2 - u^2v + \frac{4}{3}uw$

(b) $a^5 + b^5 + c^5 = \frac{1}{6}u^5 + \frac{5}{6}u^2w - \frac{5}{6}u^3v + \frac{5}{6}vw$

1.9   End of unit exercises

The following exercises, which are of varying length and difficulty, are designed to bring together many of the Maple commands that you have encountered in this unit. You should work through at least the first two of these; the last three are somewhat challenging, and you may like to reserve them for later consolidation.

Exercise 1.55

In Subsection 1.4.3 you were introduced to the hyperbolic functions sinh and cosh. In this exercise you will use Maple to learn a little more about these functions, in particular how identities involving sinh and cosh are related to identities involving sin and cos. First, recall the following identities for sin and cos:

$$\cos^2 x - \sin^2 x = \cos(2x), \tag{1.6}$$
$$\cos^2 x + \sin^2 x = 1, \tag{1.7}$$
$$\sin(2x) = 2\sin x \cos x, \tag{1.8}$$
$$\cos(2x) = 2\cos^2 x - 1, \tag{1.9}$$
$$\cos(x + y) = \cos x \cos y - \sin x \sin y \tag{1.10}$$
$$\sin(x + y) = \sin x \cos y + \sin y \cos x. \tag{1.11}$$

You are now going to use Maple to find analogous results for sinh and cosh.

As a start, type

```
> simplify(cosh(x)^2-sinh(x)^2);
```
$$1$$

This verifies the identity

$$\cosh^2 x - \sinh^2 x = 1, \tag{1.12}$$

which can be proved directly from the definitions for $\sinh x$ and $\cosh x$ on page 25, that is, from (1.2) and (1.3).

Contrast (1.12) with (1.6) and (1.7) above. Using `simplify`, `combine` or `expand` as appropriate, try to find a total of six results analogous to Equations (1.6) to (1.11). Can you spot any pattern?

Exercise 1.56

Consider the general cubic polynomial

$$f(x) = \tfrac{1}{3}ax^3 + bx^2 + cx + d,$$

where a, b, c and d are real constants. Assume that $b^2 > ac$, so that there are two stationary points of f, at x_1 and x_2. Use Maple to find the value of the constant p such that

$$f(x_1) - f(x_2) = p(x_1 - x_2)^3.$$

Exercise 1.57

Let f be a real-valued function with period 2π. Recall that the Fourier series for f of order N is defined to be

$$F_N(x) = a_0 + \sum_{n=1}^{N} \left(a_n \cos(nx) + b_n \sin(nx) \right),$$

where, for $n = 1, 2, \ldots,$

$$a_n = \frac{1}{\pi} \int_{-\pi}^{\pi} f(x) \cos(nx)\, dx, \quad b_n = \frac{1}{\pi} \int_{-\pi}^{\pi} f(x) \sin(nx)\, dx,$$

and

$$a_0 = \frac{1}{2\pi} \int_{-\pi}^{\pi} f(x)\, dx.$$

Now specifically let f be the 2π-periodic function that is equal to x^2 in the interval $-\pi < x \leq \pi$.

(a) Use Maple to find expressions for a_n, b_n and a_0.

(b) Hence find the Fourier series for f of order 2 (that is, up to and including the term in $\cos(2x)$). What is the Fourier series of order 5?

(c) Plot, on the same diagram, the graphs of f and its Fourier series of orders 2, 5 and 10.

Exercise 1.58

The function sinc x, illustrated in Figure 1.4, is usually defined as

$$\text{sinc } x = \begin{cases} \dfrac{\sin x}{x} & \text{if } x \neq 0, \\ 1 & \text{if } x = 0. \end{cases}$$

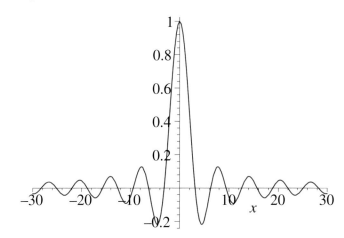

Figure 1.4   Graph of $y = \text{sinc } x$

(a) Assign the expression `sin(x)/x` to the variable `f` in Maple, and then plot `f` for $-30 \leq x \leq 30$. (Note that Maple has no problem with plotting $\dfrac{\sin 0}{0}$. However, you should satisfy yourself that the graph in Figure 1.4 looks correct at $x = 0$, given the above definition.)

There are many interesting and unusual properties of the sinc function, and you are asked to investigate two of these properties here.

(b) First, it can be shown that each maximum or minimum of the graph of sinc x corresponds to a point of intersection of the graphs of sinc x and $\cos x$. Use Maple to illustrate this, by drawing the graphs of sinc x and $\cos x$ on the same plot for $-10 \leq x \leq 10$. Then use `fsolve` to find numerical approximations for the x-coordinates of all stationary points of sinc x for $0 < x \leq 10$, and verify that these are indeed where the two graphs that you have just drawn meet.

You may wish to verify this by checking that if $f(x) = \dfrac{\sin x}{x}$, then $f'(x) = \dfrac{\cos x - f(x)}{x}$.

(c) Use Maple to evaluate the three integrals

$$\int_0^\infty \text{sinc } x \, dx, \quad \int_0^\infty \text{sinc } x \, \text{sinc } \frac{x}{3} \, dx, \quad \int_0^\infty \text{sinc } x \, \text{sinc } \frac{x}{3} \, \text{sinc } \frac{x}{5} \, dx.$$

What do you think the value of $\displaystyle\int_0^\infty \prod_{k=0}^{5} \text{sinc}(x/(2k+1)) \, dx$ is?

Use Maple to verify your conjecture.

Hint: Since `f` is assigned to an expression, not a function, you might want to use the `eval` command for sinc $\frac{x}{3}$ and sinc $\frac{x}{5}$.

Exercise 1.59

In this extended exercise you are asked to use Maple to investigate some rational bounds on π.

Before you start, you may find it useful to know that the `max` and `min` commands find the maximum and the minimum of a sequence. For example, try typing `max(2,3,4);` and `min(-1,-2,3);` at the prompt.

(a) Verify that

$$\int_0^1 \frac{x^4(1-x)^4}{1+x^2}\, dx = \tfrac{22}{7} - \pi. \tag{1.13}$$

This is in itself an interesting result, as $\frac{22}{7}$ is often used as a rational approximation for π. However, this result can also be used to obtain bounds on π in the form $a/b < \pi < c/d$, where a, b, c and d are integers.

(b) By means of

```
>   assume(x>0,x<1):
>   is(1/(1+x^2)>1/2);
>   is(1/(1+x^2)<1);
```

The command `is` is new; it should be self-explanatory, but type `?is` if you would like further information about it.

(where the output has not been printed here), verify that

$$\frac{1}{2}\int_0^1 x^4(1-x)^4\, dx < \int_0^1 \frac{x^4(1-x)^4}{1+x^2}\, dx < \int_0^1 x^4(1-x)^4\, dx.$$

(c) Evaluate $J = \int_0^1 x^4(1-x)^4\, dx$, and hence deduce that

$$\frac{1979}{630} < \pi < \frac{3959}{1260}. \tag{1.14}$$

This gives rational bounds on π. You can now use Maple to obtain tighter rational bounds.

It can be shown that the identity (1.13) can be generalized by replacing the powers of 4 by powers of any integer multiple of 4 in the following manner:

$$\int_0^1 \frac{x^{4n}(1-x)^{4n}}{2^{2(n-1)}(1+x^2)}\, dx = (-1)^n(\pi - R_n),$$

where $n = 1, 2, 3, \ldots$ and R_n is a rational number. (In identity (1.13), $n = 1$ and $R_1 = 22/7$.) Maple can verify this for any given n, and the corresponding value of R_n can be computed. Then, following the procedure of parts (b) and (c) above, new bounds on π can be found.

(d) Use Maple to calculate R_5.

(e) Hence find the upper and lower bounds on π for the case when $n = 5$. Evaluate these both as rational numbers and in decimal form, to thirty significant figures.

It can be shown that the higher the value of n, the tighter the bounds on π, but this is not considered here.

Summary of Unit 1

The main commands introduced in this unit are as follows.

`eval`   for evaluating expressions at a given value
`evalf`   for evaluating expressions to a given number of s.f.
`evalc`   for expressing complex numbers in canonical form
`plot`   for plotting simple line graphs
`solve`   for solving equations symbolically
`fsolve`   for solving equations numerically
`diff`   for differentiating expressions
`int`   for integrating expressions
`Int`   for evaluating definite integrals to a given number of s.f.
`assume`   for telling Maple to make assumptions restricting the range of expressions
`sum`   for evaluating indefinite sums symbolically
`add`   for evaluating definite sums
`expand`   for manipulating expressions
`factor`   for manipulating expressions
`simplify` for manipulating expressions
`combine`   for manipulating expressions

Learning outcomes

After studying this unit you should be able to:

- understand the use of the following symbols in Maple: `:`, `;`, `:=`, `%`, `$`, `::`, `#`;
- assign an expression to a variable;
- use `eval` to evaluate an expression at a value;
- use `evalf` to evaluate an expression to a required number of significant figures;
- solve simple equations symbolically using `solve`;
- find numerical solutions to one or more equations using `fsolve`;
- plot simple line graphs using `plot`, and use optional arguments such as `colour` and `legend` to alter the appearance of graphs;
- use `diff` to differentiate functions of one or more variables, defined explicitly or implicitly;
- use `int` to evaluate definite and indefinite integrals;
- understand the difference between `sum` and `add`, and use them both to evaluate sums;
- manipulate the output obtained from Maple using `expand`, `factor`, `simplify` and `combine`;
- use the Maple help system to find out more about any given command.

Solutions to Exercises

Remember that if you are unable to reproduce any of these solutions, then you may need to type `restart:` at the prompt before proceeding (as described on page 12). Indeed, it is recommended that the solution to each exercise starts with `restart:`.

Solution 1.1

(a) Typing

```
> 19*99; 3^20; 25!; 3/13-26/27; 2^(2^2);
```
$$1881$$
$$3486784401$$
$$15511210043330985984000000$$
$$\frac{-257}{351}$$
$$16$$

shows that $19 \times 99 = 1881$, $3^{20} = 3\,486\,784\,401$, $25! = 15\,511\,210\,043\,330\,985\,984\,000\,000$, $\frac{3}{13} - \frac{26}{27} = -\frac{257}{351}$ and $2^{(2^2)} = 16$.

(b) Similarly, typing

```
> 21.0/23; 17.0^(1/4); 1./99!; 0.216^(100);
```
$$0.9130434783$$
$$2.030543185$$
$$0.1071510288 \ 10^{-155}$$
$$0.2788528677 \ 10^{-66}$$

gives $\frac{21}{23} = 0.913\,043\,478\,3$, $17^{1/4} = 2.030\,543\,185$, $\frac{1}{99!} = 0.107\,151\,028\,8 \times 10^{-155}$ and $0.216^{100} = 0.278\,852\,867\,7 \times 10^{-66}$.

Note that the final two answers are so small that they are output in 'floating-point' form (although Maple omits the multiplication sign preceding each '10').

Solution 1.2

Typing

```
> (5-8*I)^2; I^(-1); (1+I)/(5+2*I);
```
$$-39 - 80\,I$$
$$-I$$
$$\frac{7}{29} + \frac{3}{29}\,I$$

```
> (4/3+2*I/5)^4;
```
$$\frac{74896}{50625} + \frac{11648}{3375}\,I$$

gives $(5-8i)^2 = -39 - 80i$, $i^{-1} = -i$, $\frac{1+i}{5+2i} = \frac{7}{29} + \frac{3}{29}i$ and $\left(\frac{4}{3} + \frac{2}{5}i\right)^4 = \frac{74\,896}{50\,625} + \frac{11\,648}{3375}i$.

Solution 1.3

The assignments are entered as follows:

```
> a := 10; b := 20; c := a+2*b;
```
$$a := 10$$
$$b := 20$$
$$c := 50$$

Then evaluate

```
> a+b+c;
```
$$80$$

so $a + b + c = 80$.

Solution 1.4

After a little experimentation, increasing `Digits` to 30 (say) shows the required repetition:

```
> Digits := 30:
> 21.0/23;
```
$$0.913043478260869565217391304348$$

so $21/23 = 0.\overline{913043478260869565217}3$. (The bar over the digits here indicates those that are to be repeated.)

Solution 1.5

You should obtain the output

```
> z;
```
$$13 + x^2$$

because when z was evaluated, a still had the value 3.

Solution 1.6

(a) The trigonometric terms are evaluated as follows:

```
> tan(Pi/3); cos(Pi/3)^2; sec(Pi/6);
```
$$\sqrt{3}$$
$$\frac{1}{4}$$
$$\frac{2\sqrt{3}}{3}$$

```
> 1+cot(Pi/4)^2; csc(Pi/4)^2;
```
$$2$$
$$2$$

The last two examples above are an illustration of the identity $\operatorname{cosec}^2 \theta = 1 + \cot^2 \theta$.

(b) In the evaluation of $\sqrt{3 + \sqrt{3 + \sqrt{3}}}$, the 3s are all input as 3.0 in order to obtain the decimal output requested:

```
> sqrt(3.0+sqrt(3.0+sqrt(3.0)));
```
$$2.274934669$$

So $\sqrt{3 + \sqrt{3 + \sqrt{3}}} = 2.274\,934\,669$ to ten significant figures.

(It is also possible just to input the 'innermost' 3 as a decimal: `sqrt(3+sqrt(3+sqrt(3.0)));`. This is because all of the evaluations of `sqrt` 'see' the innermost 3.)

Finally, typing

```
> exp(3*ln(4));
```
$$64$$

gives $e^{3\ln 4} = 64$, as expected.

Solution 1.7

Typing `?ithprime` brings up the required help page. On typing

```
>  ithprime(100); ithprime(1000);
```

$$541$$
$$7919$$

```
>  ithprime(10000);
```

$$104729$$

it can be seen that the 100th prime is 541, the 1000th is 7919, and the 10 000th is 104 729. Note that in the input, 1000 and 10000 must be typed without spaces.

Solution 1.8

(a) Typing `?sin`, `?arcsin` or `?invtrig` will all bring up useful help pages. There are several ways of evaluating $\arcsin(1/2)$, the most straightforward being

```
>  arcsin(1/2);
```

$$\frac{\pi}{6}$$

so that, as expected, the result $\arcsin(1/2) = \pi/6$ is obtained.

(b) Typing

```
>  sec(arctan(x));
```

$$\sqrt{1+x^2}$$

shows that $\sec(\arctan(x)) = \sqrt{1+x^2}$, which comes from the identity $\sec^2\theta = 1 + \tan^2\theta$ (or from the definitions of sec and arctan).

Solution 1.9

As in the text, in each case the expression to be evaluated is assigned to the name y first. This is partly so that the input can be checked, but it is also good practice, because two short command lines are easier to work with than one long one.

(a)

```
>  y := x^3-3*x^2+2*x-1;
```

$$y := x^3 - 3x^2 + 2x - 1$$

```
>  eval(y,x=5);
```

$$59$$

So when $x = 5$ the given expression takes the value 59.

(b)

```
>  y := sin(x)*cos(x)^3;
```

$$y := \sin(x)\cos(x)^3$$

```
>  eval(y,x=Pi/4);
```

$$\frac{1}{4}$$

So when $x = \pi/4$ the given expression takes the value 1/4.

(c)

```
>  y := exp(t);
```

$$y := e^t$$

```
>  eval(y,t=1);
```

$$e$$

So when $t = 1$ the given expression takes the value e.

(d)

```
>  y := ln(u+1+sqrt(u^2-3));
```

$$y := \ln(u + 1 + \sqrt{u^2 - 3})$$

```
>  eval(y,u=2);
```

$$2\ln(2)$$

So when $u = 2$ the given expression takes the value $2\ln 2$ (or $\ln 4$).

(e)

```
>  y := 3*sin(2*t)+4*cos(3*t);
```

$$y := 3\sin(2t) + 4\cos(3t)$$

```
>  eval(x,t=Pi/3);
```

$$\frac{3\sqrt{3}}{2} - 4$$

So when $t = \pi/3$ the given expression takes the value $\frac{3\sqrt{3}}{2} - 4$.

Solution 1.10

First assign the given expressions to π_1 and π_2, and then compute the absolute errors $|\pi_1 - \pi|$ and $|\pi_2 - \pi|$ (after a little experimentation to find the number of digits to use with `evalf`):

```
>  pi1 := 12/sqrt(190)
    *ln((2*sqrt(2)+sqrt(10))*(3+sqrt(10)));
```

$$\pi 1 := \frac{6}{95}\sqrt{190}\ln((2\sqrt{2} + \sqrt{10})(3 + \sqrt{10}))$$

```
>  pi2 := (9^2+19^2/22)^(1/4);
```

$$\pi 2 := \frac{2143^{(1/4)}\, 22^{(3/4)}}{22}$$

```
>  evalf[20](abs(pi1-Pi));
```

$$0.3 \ 10^{-18}$$

```
>  evalf[20](abs(pi2-Pi));
```

$$0.10071471132 \ 10^{-8}$$

Hence π_1 is accurate to 18 decimal places, whilst π_2 is accurate to only 8. (You may recall that if the absolute error is less than 0.5×10^{-n}, then the approximate value is accurate to n decimal places.)

Solution 1.11

From de Moivre's theorem and

```
>  z := cos(theta)+I*sin(theta);
```

$$z := \cos(\theta) + \sin(\theta)\,I$$

```
>  evalc(Re(z^2));
```

$$\cos(\theta)^2 - \sin(\theta)^2$$

```
>  evalc(Im(z^2));
```

$$2\cos(\theta)\sin(\theta)$$

we obtain the standard results $\cos(2\theta) = \cos^2\theta - \sin^2\theta$ and $\sin(2\theta) = 2\sin\theta\cos\theta$.

Solution 1.12

Type

```
>  z := 3/2*ln(2)+I*Pi/4;
```
$$z := \frac{3}{2}\ln(2) + \frac{1}{4}I\pi$$

```
>  exp(z);
```
$$\mathbf{e}^{\left(\frac{3}{2}\ln(2)+\frac{1}{4}I\pi\right)}$$

```
>  evalc(%);
```
$$2 + 2I$$

so $e^z = 2 + 2i$.

Solution 1.13

Figure 1.5 is obtained with the commands

```
>  y := sin(10*cos(x)):
>  plot(y,x=-Pi..Pi);
```

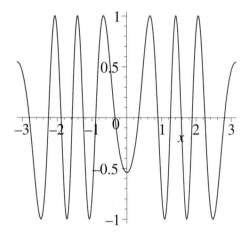

Figure 1.5

Solution 1.14

The following command produces Figure 1.6:

```
>  plot([1,1+x,1+x+x^2/2!,1+x+x^2/2!+x^3/3!,
   exp(x)],x=0..1);
```

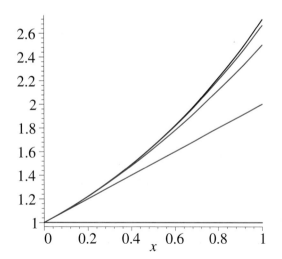

Figure 1.6

The graphs of the first four functions (red on this figure) are progressively better approximations to the graph of $y = e^x$ (black on the figure), as expected. (To convince yourself of this, use the `legend` option within `plot`, as described in the text.)

Solution 1.15

After a little experimentation, the roots are seen to lie in the interval $[-1, 5]$, so the following commands are used:

```
>  a := sin(x);
```
$$a := \sin(x)$$

```
>  b := x^3-5*x^2+4;
```
$$b := x^3 - 5x^2 + 4$$

```
>  plot([a,b],x=-1..5);
```

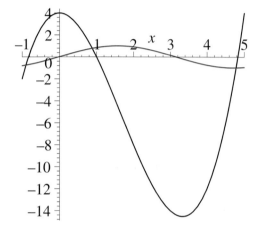

Figure 1.7

The x-values of the cursor coordinates at the points of intersection are approximately -0.90, 0.89 and 4.78.

Solution 1.16

The command

```
>  plot([tan(x),1/x],x=0..15,0..2,
   discont=true);
```

produces Figure 1.8.

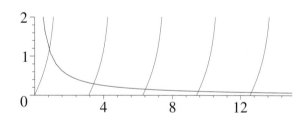

Figure 1.8

The x-values of the cursor coordinates at the points of intersection are (to one decimal place) 0.9, 3.4, 6.4, 9.5 and 12.6, and these are the required roots.

Solution 1.17

(a) Type

```
>  solve(x^2-x-2025=0,x);
```
$$\frac{1}{2} + \frac{\sqrt{8101}}{2}, \; \frac{1}{2} - \frac{\sqrt{8101}}{2}$$

The solutions of $x^2 - x - 2025 = 0$ are $x = (1 \pm \sqrt{8101})/2$.

(b) Type

```
> solve(x^3-6*x^2-19*x+24=0,x);
```
$$1, -3, 8$$

The solutions of $x^3 - 6x^2 - 19x + 24 = 0$ are $x = 1$, -3 and 8.

(c) As the equation to be solved is quite lengthy, assign it first:

```
> eq := 2*x^4-11*x^3-20*x^2+113*x+60=0;
```
$$eq := 2x^4 - 11x^3 - 20x^2 + 113x + 60 = 0$$

Then solve it:

```
> solve(eq,x);
```
$$\frac{-1}{2}, -3, 4, 5$$

The solutions of $2x^4 - 11x^3 - 20x^2 + 113x + 60 = 0$ are $x = -1/2$, -3, 4 and 5.

Solution 1.18

First input the equations, and then solve them:

```
> eq := {a*x+b*y=A,c*x+d*y=B};
```
$$eq := \{ax + by = A, \, cx + dy = B\}$$

```
> sol := solve(eq,{x,y});
```
$$sol := \{y = \frac{aB - cA}{ad - cb}, \, x = -\frac{bB - Ad}{ad - cb}\}$$

So the point of intersection is
$$(x, y) = \left(\frac{Ad - bB}{ad - cb}, \frac{aB - cA}{ad - cb} \right).$$

Now assign x and y these values and compute the required distance:

```
> assign(sol);
```
```
> sqrt(x^2+y^2);
```
$$\sqrt{\frac{(bB - Ad)^2}{(ad - cb)^2} + \frac{(aB - cA)^2}{(ad - cb)^2}}$$

This expression represents the required distance.

Solution 1.19

Using the command

```
> fsolve(x*sin(x)=1/2,x=0..2);
```

and then altering the range of x-values as appropriate, the solutions are

(a) 0.740 840 955 1,
(b) 2.972 585 490,
(c) 6.361 859 813,
(d) 9.371 398 787

(to ten significant figures).

From Figure 1.9, obtained using the command

```
> plot([x*sin(x),1/2],x=0..10);
```

it can be seen that there are no roots in the range $4 < x < 6$.

(You could equally well have used the command `plot([x*sin(x),1/2],x=4..6)` to see that there are no solutions in the range $4 < x < 6$.)

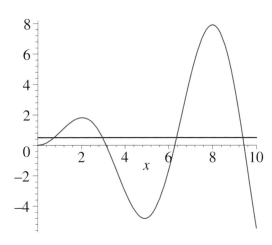

Figure 1.9

Solution 1.20

Using the command

```
> fsolve(x*tan(x)=1,x=0..1);
```

and then altering the range of x-values as appropriate, the solutions (to six decimal places) are $x = 0.860 334$, 3.425 618, 6.437 298, 9.529 334 and 12.645 287.

Solution 1.21

(a) Using the command

```
> plot(x^3+3*x^2-2*x+1,x=-4..2);
```

produces the graph shown in Figure 1.10, from which it can be seen that there is only one real root.

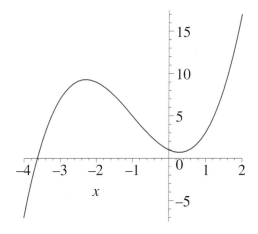

Figure 1.10

The command

```
> fsolve(x^3+3*x^2-2*x+1=0,x=-4..-3);
```

gives the root as $-3.627 365$, when rounded to six decimal places.

(b) Typing

```
> solve(x^3+3*x^2-2*x+1=0,x);
```

does explicitly give the three roots (one real, two complex), but not in a very usable form (we have not printed the output here). However, the command

```
> evalf(solve(x^3+3*x^2-2*x+1=0,x));
```

gives the approximate values of the three roots as $-3.627 365$ and $0.313 683 \pm 0.421 053i$.

Solution 1.22

(a) The command

```
> plot([sinh(x),cosh(x)],x=-2..2,
  legend=["sinh","cosh"]);
```

produces Figure 1.11.

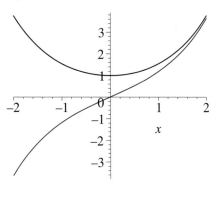

Figure 1.11

The graph of $\sinh x$ is increasing for all x, with no stationary points, whereas $\cosh x$ has one stationary point, a minimum at $(0,1)$. As $x \to \infty$, both graphs tend to the graph of $e^x/2$; this follows from the definitions (1.2) and (1.3), and the fact that $e^{-x} \to 0$. Similarly, as $x \to -\infty$, the graphs of $\sinh x$ and $\cosh x$ tend to the graphs of $-e^{-x}/2$ and $e^{-x}/2$, respectively.

The function $\sinh x$ is an odd function (that is, $\sinh(-x) = -\sinh(x)$, as is immediate from its definition); its graph thus has rotational symmetry about the origin.

The function $\cosh x$ is an even function (that is, $\cosh(-x) = \cosh(x)$, again immediate from the definition); its graph is thus symmetrical about the y-axis.

(b) To plot the graph of $\tanh x$ for $-5 \le x \le 5$, use the command

```
> plot(tanh(x),x=-5..5);
```

which produces Figure 1.12.

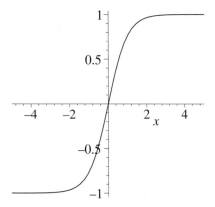

Figure 1.12

This is consistent with $\tanh x$ being the ratio $\dfrac{\sinh x}{\cosh x}$ as given in definition (1.4). In particular, as $x \to \infty$, $\tanh x \to 1$, and as $x \to -\infty$, $\tanh x \to -1$. Also, as $\sinh x$ is odd and $\cosh x$ is even, $\tanh x$ is odd.

(c) Evaluate $\tanh 5$ to 15 s.f. using `evalf` as follows:

```
> evalf[15](tanh(5));
```
$$0.999909204262595$$

and the others follow in a similar manner:

```
> evalf[15](tanh(10));
```
$$0.999999995877693$$

```
> evalf[15](tanh(15));
```
$$0.999999999999813$$

```
> evalf[15](tanh(20));
```
$$1.$$

These verify what is expected from the definition of $\tanh x$, and what was seen in Figure 1.12, namely that $\tanh x$ approaches the value 1 for large x (where 20 is not even that large!).

(d) Typing

```
> solve(12*cosh(x)^2+7*sinh(x)=24,x);
```
$$-\ln(2) + \pi I, \ \ln(2), \ \ln(3) + \pi I, \ -\ln(3)$$

shows that there are two real solutions, $x = \ln 2$ and $x = -\ln 3$.

Solution 1.23

(a) Start by defining y and then plot its graph over the given range:

```
> y := 12*x^5-15*x^4+20*x^3-330*x^2+600*x+2;
```
$$y := 12\,x^5 - 15\,x^4 + 20\,x^3 - 330\,x^2 + 600\,x + 2$$

```
> plot(y,x=-1..3);
```

The graph is shown in Figure 1.13.

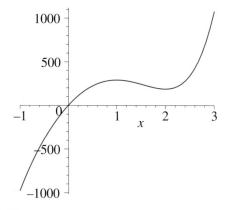

Figure 1.13

Only two stationary points are shown on the graph, and we would expect four for a quintic. However, two may be complex. To find the coordinates of the stationary points, type

```
> solve(diff(y,x)=0,x);
```
$$1, \ 2, \ -1 + 2\,I, \ -1 - 2\,I$$

and indeed there are only two real values of x for which $dy/dx = 0$. To find the corresponding y-values, type

```
> eval(y,x=1); eval(y,x=2);
```
$$289$$
$$186$$

So there are (real) stationary points at $(1, 289)$ and $(2, 186)$.

(b) The following commands produce Figure 1.14:

```
> y := sin(x^cos(x));
```
$$y := \sin(x^{\cos(x)})$$

```
> plot(y,x=0..2*Pi);
```

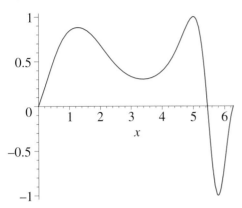

Figure 1.14

Since there are infinitely many stationary points unless we restrict the domain, we need to use `fsolve` to find approximate numerical solutions to $dy/dx = 0$ within the given interval, and we can use Figure 1.14 to identify suitable subintervals within which `fsolve` can search. So, for example, the commands

```
> fsolve(diff(y,x)=0,x=1..2);
```
$$1.272850698$$

```
> eval(y,x=%);
```
$$0.8788238014$$

show that one stationary point is at approximately $(1.272\,851, 0.878\,824)$. In a similar fashion, the other stationary points are found to be at $(3.379\,916, 0.301\,463)$, $(4.996\,904, 1)$ and $(5.793\,406, -1)$. Placing the cursor over the graph and left-clicking the mouse verifies these values (to a fair degree of accuracy).

Solution 1.24

First define the expressions as follows:

```
> y1 := x^x;  y2 := x^y1;  y3 := x^y2;
```
$$y1 := x^x$$
$$y2 := x^{(x^x)}$$
$$y3 := x^{(x^{(x^x)})}$$

Then plot them using

```
> plot([y1,y2,y3],x=0..1,
    legend=["y1","y2","y3"]);
```

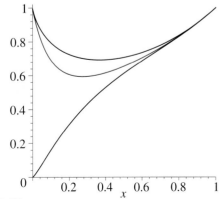

Figure 1.15

(To save space, we have not given the legend in Figure 1.15.)

Then typing

```
> solve(diff(y1,x)=0,x);
```
$$\frac{1}{e}$$

shows that y_1 has a stationary point at $x = 1/e$, which must be a minimum from its graph.

Typing

```
> fsolve(diff(y3,x)=0,x=0..1);
```
$$0.2746893853$$

shows that y_3 has a stationary point at 0.2747 (to four d.p.) and again this is a minimum from the graph.

Solution 1.25

Use the $AC - B^2$ test to classify the stationary point. First define f by typing

```
> f := 2*x^2-x*y-3*y^2-3*x+7*y-(x-1)^3;
```
$$f := 2\,x^2 - x\,y - 3\,y^2 - 3\,x + 7\,y - (x-1)^3$$

and then solve $f_x = f_y = 0$:

```
> solve({diff(f,x)=0,diff(f,y)=0},{x,y});
```
$$\{x = 1, y = 1\}, \{x = \frac{43}{18}, y = \frac{83}{108}\}$$

So there are stationary points at $(1, 1)$ and $(43/18, 83/108)$. To classify the stationary point at $(1, 1)$, find A, B and C by typing

```
> A := diff(f,x$2):
> B := diff(f,x,y):
> C := diff(f,y$2):
```

and evaluate $AC - B^2$ at $(1, 1)$:

```
> eval(A*C-B^2,{x=1,y=1});
```
$$-25$$

As $AC - B^2 < 0$, there is a saddle at $(1, 1)$.

Solution 1.26

Having defined `z`, we can use `solve` all in one go:

```
> z := ln(x^2+y^2);
```
$$z := \ln(x^2 + y^2)$$

```
> solve((diff(z,x))^2+(diff(z,y))^2
    =a*exp(-z),a);
```
$$4$$

Hence $a = 4$.

Solution 1.27

Proceed as follows:

```
> y := sin(x+x^2);
```
$$y := \sin(x + x^2)$$

```
> y0 := eval(y,x=0);
```
$$y0 := 0$$

```
> y1 := eval(diff(y,x),x=0);
```
$$y1 := 1$$

```
> y2 := eval(diff(y,x$2),x=0);
```
$$y2 := 2$$

```
>  y3 := eval(diff(y,x$3),x=0);
```
$$y3 := -1$$

so the Taylor series about $x = 0$ is
$$x + x^2 - \tfrac{1}{6}x^3 + \cdots.$$

Similarly, from the commands
```
>  yp0 := eval(y,x=Pi);
```
$$yp0 := -\sin(\pi^2)$$
```
>  yp1 := eval(diff(y,x),x=Pi);
```
$$yp1 := -\cos(\pi^2)\,(1 + 2\pi)$$
```
>  yp2 := eval(diff(y,x$2),x=Pi);
```
$$yp2 := \sin(\pi^2)\,(1 + 2\pi)^2 - 2\cos(\pi^2)$$

the Taylor series about $x = \pi$ is
$$-\sin(\pi^2) - \cos(\pi^2)(1 + 2\pi)(x - \pi)$$
$$+ (\tfrac{1}{2}\sin(\pi^2)(1 + 2\pi)^2 - \cos(\pi^2))(x - \pi)^2 + \cdots.$$

Solution 1.28

In a manner similar to the previous exercise, start by defining f as follows:
```
>  f := ln((1+sin(x))/cos(x))
   -2*sin(x)+x*cos(x);
```
$$f := \ln\left(\frac{1 + \sin(x)}{\cos(x)}\right) - 2\sin(x) + x\cos(x)$$

and then evaluate f and its derivatives at $x = 0$:
```
>  eval(f,x=0);
```
$$0$$
```
>  eval(diff(f,x),x=0);
```
$$0$$
```
>  eval(diff(f,x$2),x=0);
```
$$0$$
```
>  eval(diff(f,x$3),x=0);
```
$$0$$
```
>  eval(diff(f,x$4),x=0);
```
$$0$$
```
>  eval(diff(f,x$5),x=0);
```
$$8$$

So the first term of the Taylor series is $f^{(5)}(0)\,x^5/5! = x^5/15$.

Solution 1.29

(a) Typing
```
>  diff(sin(f(x)),x);
```
$$\cos(f(x))\left(\frac{d}{dx}\,f(x)\right)$$

shows that the derivative of $\sin(f(x))$ is $f'(x)\cos(f(x))$.

(b) Similarly, typing
```
>  diff(sin(exp(f(x))),x);
```
$$\cos(e^{f(x)})\left(\frac{d}{dx}\,f(x)\right)e^{f(x)}$$

shows that the derivative of $\sin(e^{f(x)})$ is $f'(x)\,e^{f(x)}\cos(e^{f(x)})$.

(c) Finally,
```
>  diff(exp(sqrt(1+f(x)*g(x))),x);
```
$$\frac{1}{2}\frac{\left(\left(\frac{d}{dx}\,f(x)\right)g(x) + f(x)\left(\frac{d}{dx}\,g(x)\right)\right)e^{(\sqrt{1+f(x)\,g(x)})}}{\sqrt{1 + f(x)\,g(x)}}$$

shows that the derivative of $\exp(\sqrt{1 + f(x)g(x)})$ is
$$\frac{f'(x)g(x) + f(x)g'(x)}{2\sqrt{1 + f(x)g(x)}}\exp(\sqrt{1 + f(x)g(x)}).$$

Solution 1.30

First define the equation to be differentiated, and then differentiate it:
```
>  eq1 := x^2+y(x)^2+3*x*y(x)=0;
```
$$eq1 := x^2 + y(x)^2 + 3\,x\,y(x) = 0$$
```
>  eq2 := diff(eq1,x);
```
$$eq2 :=$$
$$2\,x + 2\,y(x)\left(\frac{d}{dx}\,y(x)\right) + 3\,y(x) + 3\,x\left(\frac{d}{dx}\,y(x)\right) = 0$$

Then solve this equation for dy/dx by typing
```
>  solve(eq2,diff(y(x),x));
```
$$-\frac{2\,x + 3\,y(x)}{2\,y(x) + 3\,x}$$

So $\dfrac{dy}{dx} = -\dfrac{2x + 3y}{2y + 3x}.$

Solution 1.31

First set up the alias by typing
```
>  alias(y=y(x));
```
$$y$$

and then proceed as in the previous exercise, but typing y instead of y(x):
```
>  eq1 := x^2+y^2+3*x*y=0;
```
$$eq1 := x^2 + y^2 + 3\,x\,y = 0$$
```
>  eq2 := diff(eq1,x);
```
$$eq2 := 2\,x + 2\,y\left(\tfrac{\partial}{\partial x}\,y\right) + 3\,y + 3\,x\left(\tfrac{\partial}{\partial x}\,y\right) = 0$$
```
>  solve(eq2,diff(y,x));
```
$$-\frac{2\,x + 3\,y}{2\,y + 3\,x}$$

which is what was found before.

Although it was not asked for in the question, say we now wish to remove the alias. Proceed as follows:
```
>  alias(y=y);
```
and check that y now just means y:
```
>  diff(y,x);
```
$$0$$

Solution 1.32

This problem might be familiar to you if you have previously studied the wave equation. One way to prove that $f(x - ct)$ satisfies $c^2 f_{xx} = f_{tt}$ is to evaluate $c^2 f_{xx} - f_{tt}$ for $f(x - ct)$ as follows:
```
>  c^2*diff(f(x-c*t),x$2)-diff(f(x-c*t),t$2);
```
$$0$$

Hence the given equation is satisfied by $f(x - ct)$.

Solution 1.33

Throughout the solutions to this exercise, the arbitrary constant of integration has been omitted.

(a) Type

```
> int(sqrt(exp(x)-1),x);
```
$$2\sqrt{e^x - 1} - 2\arctan(\sqrt{e^x - 1})$$

So
$$\int \sqrt{e^x - 1}\,dx = 2\left(\sqrt{e^x - 1} - \arctan(\sqrt{e^x - 1})\right).$$

(b) Typing

```
> int(x^2*(a*x+b)^(5/2),x);
```
$$\frac{2\,(a\,x + b)^{(7/2)}\,(8\,b^2 - 28\,b\,x\,a + 63\,x^2\,a^2)}{693\,a^3}$$

shows that
$$\int x^2(ax+b)^{5/2}\,dx$$
$$= \frac{2(ax+b)^{7/2}(8b^2 - 28bxa + 63x^2a^2)}{693a^3}.$$

(c) In this answer the Maple output is too long to be included here, but typing

```
> int(sinh(6*x)*(sinh(x))^4,x);
```

gives
$$\int \sinh(6x)\sinh^4 x\,dx = \tfrac{1}{16}\cosh(6x) + \tfrac{1}{160}\cosh(10x)$$
$$+ \tfrac{1}{32}\cosh(2x) - \tfrac{1}{32}\cosh(8x) - \tfrac{1}{16}\cosh(4x).$$

(d) Type

```
> int((cosh(x))^(-6),x);
```
$$\frac{1}{5}\frac{\sinh(x)}{\cosh(x)^5} + \frac{4}{15}\frac{\sinh(x)}{\cosh(x)^3} + \frac{8}{15}\frac{\sinh(x)}{\cosh(x)}$$

to obtain the result
$$\int (\cosh x)^{-6}\,dx = \frac{\sinh x}{5\cosh^5 x} + \frac{4\sinh x}{15\cosh^3 x} + \frac{8\sinh x}{15\cosh x}.$$

(e) Finally, type

```
> int(sin(ln(x)),x);
```
$$-\frac{1}{2}\cos(\ln(x))\,x + \frac{1}{2}\sin(\ln(x))\,x$$

to obtain the result
$$\int \sin(\ln x)\,dx = -\tfrac{1}{2}x\cos(\ln x) + \tfrac{1}{2}x\sin(\ln x).$$

Solution 1.34

(a) The following evaluates the given definite integral:

```
> int(1/(1+x^3),x=1/2..1);
```
$$-\frac{1}{6}\ln(3) + \frac{1}{3}\ln(2) + \frac{\sqrt{3}\,\pi}{18}$$

So
$$\int_{1/2}^1 \frac{1}{1+x^3}\,dx = -\frac{1}{6}\ln 3 + \frac{1}{3}\ln 2 + \frac{\sqrt{3}}{18}\pi.$$

(b) To evaluate the integral, type

```
> int(x^2*arctan(x),x=0..1);
```
$$\frac{\pi}{12} - \frac{1}{6} + \frac{1}{6}\ln(2)$$

So
$$\int_0^1 x^2\arctan x\,dx = \frac{\pi}{12} - \frac{1}{6} + \frac{1}{6}\ln 2.$$

(c) To evaluate the integral, type

```
> int(1/((1+x)*(1+x^2)),x=0..infinity);
```
$$\frac{\pi}{4}$$

So
$$\int_0^\infty \frac{1}{(1+x)(1+x^2)}\,dx = \frac{\pi}{4}.$$

Solution 1.35

(a) Typing

```
> evalf[6](Int(exp(x^3),x=0..1));
```
$$1.34190$$

gives $\int_0^1 e^{x^3}\,dx = 1.341\,90$ to six significant figures.

(b) Typing

```
> evalf[6](Int(1/sqrt(1+x^4),x=0..10));
```
$$1.75408$$

gives $\int_0^{10} \frac{1}{\sqrt{1+x^4}}\,dx = 1.754\,08$ to six significant figures. Note, however, that in this example `int` will produce the answer far faster than `Int` (try it with 200 significant figures, for example, to really see the difference). This is because Maple *can* evaluate the integral symbolically, and it is quicker for it to do so, and then substitute the limits in.

(c) Typing

```
> evalf[6](Int(sin(exp(x/2)),x=0..5));
```
$$1.10398$$

gives $\int_0^5 \sin(e^{x/2})\,dx = 1.103\,98$ to six significant figures.

Solution 1.36

(a) The following commands evaluate the integral for $n > 0$:

```
> assume(n>0): int(1/x,x=1..n);
```
$$\ln(n\sim)$$

So $\int_1^n \frac{1}{x}\,dx = \ln n$, for $n > 0$.

(b) Similarly, typing

```
> assume(a<0):
> int(exp(a*x^2),x=-infinity..infinity);
```
$$\frac{\sqrt{\pi}}{\sqrt{-a\sim}}$$

shows that $\int_{-\infty}^\infty e^{ax^2}\,dx = \sqrt{\dfrac{\pi}{-a}}$, for $a < 0$.

Solution 1.37

(a) With this definite sum, use `add`:

```
>   add(r^2+3*r-2,r=1..10);
                            530
```

So $\sum_{r=1}^{10} (r^2 + 3r - 2) = 530$.

(b) Similarly,

```
>   add(2^i,i=1..50);
                      2251799813685246
```

gives $\sum_{i=1}^{50} 2^i = 2\,251\,799\,813\,685\,246$.

(c) This is an indefinite sum, so use `sum`:

```
>   sum(a^k,k=0..n-1);
```
$$\frac{a^n}{a-1} - \frac{1}{a-1}$$

Arranged as $\sum_{k=0}^{n-1} a^k = \frac{a^n - 1}{a - 1}$, this is the standard expression for the sum of a finite geometric series whose first term is 1 and common ratio is a.

(d) Finally, use `sum` again for this infinite sum:

```
>   sum(1/k!,k=0..infinity);
                             e
```

So $\sum_{k=0}^{\infty} \frac{1}{k!} = e$, a result that might be familiar to you.

Solution 1.38

For each of these infinite sums, use `sum`.

(a)

```
>   sum(2/((n+1)*(n+2)),n=1..infinity);
                             1
```

So $\sum_{n=1}^{\infty} \frac{2}{(n+1)(n+2)} = 1$, as required.

(b)

```
>   sum((k^2+k-1)/(k+2)!,k=0..infinity);
                             0
```

So $\sum_{k=0}^{\infty} \frac{k^2 + k - 1}{(k+2)!} = 0$, as required.

(c)

```
>   sum(1/k^2,k=1..infinity);
```
$$\frac{\pi^2}{6}$$

So $\sum_{k=1}^{\infty} \frac{1}{k^2} = \frac{\pi^2}{6}$, as required.

Solution 1.39

In both of these, use the `add(evalf(...))` construction.

(a)

```
>   add(evalf(exp(-sqrt(k))),k=1..10);
                       1.338642283
```

So $\sum_{k=1}^{10} e^{-\sqrt{k}} = 1.338\,642\,283$ to ten significant figures.

(b)

```
>   add(evalf(1/sqrt(k)),k=1..100);
                       18.58960383
```

So $\sum_{k=1}^{100} \frac{1}{\sqrt{k}} = 18.589\,603\,83$ to ten significant figures.

Solution 1.40

(a) The commands

```
>   z := (2*n+1)*exp(m):
>   add(add(evalf(z),n=0..10),m=0..10);
                     0.4216217635 10^7
```

show the solution to be $4.216\,217\,635 \times 10^6$ to ten significant figures.

(b) Similarly, the second double sum is evaluated using the commands

```
>   z := 1/(n^2+ln(22+m)):
>   add(add(evalf(z),m=-10..10),n=-20..20);
                       35.86770332
```

which show the solution to be $35.867\,703\,32$ to ten significant figures.

Solution 1.41

In both cases it is better to use `mul`, rather than `product`, because `mul` is designed to evaluate *definite* products.

(a)

```
>   mul(r^2,r=1..10);
                     13168189440000
```

So $\prod_{r=1}^{10} r^2 = 13\,168\,189\,440\,000$.

(b)

```
>   mul(1-q^r,r=1..5);
```
$$(1-q)(1-q^2)(1-q^3)(1-q^4)(1-q^5)$$

So
$$\prod_{r=1}^{5}(1-q^r) = (1-q)(1-q^2)(1-q^3)(1-q^4)(1-q^5),$$

as expected.

Solution 1.42

(a) The first expansion is a straightforward application of `expand`:

```
>   expand(sin(3*x));
```
$$4\sin(x)\cos(x)^2 - \sin(x)$$

(b) For this partial expansion, proceed as in the text:

```
>   s := expand(sin(x+y));
```
$$s := \sin(x)\cos(y) + \cos(x)\sin(y)$$

```
>   eval(s,y=2*x);
```
$$\sin(x)\cos(2x) + \cos(x)\sin(2x)$$

Solution 1.43

First define f and then expand:

```
>  f := (x+y+1)^2;
```
$$f := (x + y + 1)^2$$

```
>  expand(f);
```
$$x^2 + 2\,x\,y + 2\,x + y^2 + 2\,y + 1$$

So $(x + y + 1)^2 = x^2 + 2xy + 2x + y^2 + 2y + 1$.

The constant 1 in this expansion is now subtracted:

```
>  factor(f-1);
```
$$(x + y + 2)\,(x + y)$$

So $f - 1 = (x + y + 2)(x + y)$.

Solution 1.44

Typing

```
>  factor(expand(sinh(5*x)/sinh(x)));
```
$$(4\cosh(x)^2 - 2\cosh(x) - 1)\,(4\cosh(x)^2 + 2\cosh(x) - 1)$$

shows that

$$\frac{\sinh 5x}{\sinh x} = (4\cosh^2 x - 2\cosh x - 1)$$
$$\times\,(4\cosh^2 x + 2\cosh x - 1).$$

Hence one solution is $p = 4$, $q = -2$, $r = -1$, $s = 4$, $t = 2$ and $u = -1$.

Solution 1.45

The results are obtained as follows:

```
>  factor(sum(r,r=1..n));
```
$$\frac{n\,(n + 1)}{2}$$

```
>  factor(sum(r^2,r=1..n));
```
$$\frac{n\,(n + 1)\,(2\,n + 1)}{6}$$

```
>  factor(sum(r^3,r=1..n));
```
$$\frac{n^2\,(n + 1)^2}{4}$$

Solution 1.46

Typing

```
>  ifactor(10!);
```
$$(2)^8\,(3)^4\,(5)^2\,(7)$$

shows that $10! = 2^8 3^4 5^2 7$. Similarly, to factorize $10! + 1$, type

```
>  ifactor(10!+1);
```
$$(11)\,(329891)$$

Hence $10! + 1 = 11 \times 329\,891$. Finally, typing

```
>  ifactor(11!+1);
```
$$(39916801)$$

shows that $11! + 1$ is prime, as it cannot be broken down into factors.

In fact, the last part can also be done by using the command `isprime` (which you will meet in *Unit 2*), as follows:

```
>  isprime(11!+1);
```
$$true$$

Solution 1.47

In each case, first use `diff` to perform the differentiation, and then use `simplify` to get the answer into the required form.

(a)

```
>  diff((1+sin(x))/(1+cos(x)),x);
```
$$\frac{\cos(x)}{1 + \cos(x)} + \frac{(1 + \sin(x))\,\sin(x)}{(1 + \cos(x))^2}$$

```
>  simplify(%);
```
$$\frac{\cos(x) + \sin(x) + 1}{1 + 2\cos(x) + \cos(x)^2}$$

(b)

```
>  diff(ln(sqrt((1+x)/(1-x))),x);
```
$$\frac{\left(\dfrac{1}{1 - x} + \dfrac{1 + x}{(1 - x)^2}\right)(1 - x)}{2\,(1 + x)}$$

```
>  simplify(%);
```
$$-\frac{1}{-1 + x^2}$$

(c)

```
>  diff(ln((1+x)^(1/2)/(1-x)^(1/3)),x);
```
$$\frac{\left(\dfrac{1}{2\sqrt{1 + x}\,(1 - x)^{(1/3)}} + \dfrac{\sqrt{1 + x}}{3\,(1 - x)^{(4/3)}}\right)(1 - x)^{(1/3)}}{\sqrt{1 + x}}$$

```
>  simplify(%);
```
$$\frac{-5 + x}{6\,(-1 + x^2)}$$

Solution 1.48

In this answer we have suppressed the intermediate output as it is too lengthy.

First, define f by typing

```
>  f := 1/sqrt((x-a)^2+(y-b)^2+(z-c)^2):
```

and then find $f_{xx} + f_{yy} + f_{zz}$:

```
>  ans := diff(f,x,x)+diff(f,y,y)
          +diff(f,z,z):
```

The output does not immediately look as though it will evaluate to zero, but typing

```
>  simplify(ans);
```
$$0$$

shows that indeed $f_{xx} + f_{yy} + f_{zz} = 0$.

Solution 1.49

Typing

```
>  simplify(arctan(tan(x)),symbolic);
```
$$x$$

produces the required result.

Solution 1.50

To obtain each required expression, use `combine` as follows.

```
> combine(sin(x)^7);
```

$$-\frac{1}{64}\sin(7\,x)+\frac{7}{64}\sin(5\,x)-\frac{21}{64}\sin(3\,x)+\frac{35}{64}\sin(x)$$

shows that

$$\sin^7 x = -\tfrac{1}{64}\sin(7x)+\tfrac{7}{64}\sin(5x)$$
$$-\tfrac{21}{64}\sin(3x)+\tfrac{35}{64}\sin x.$$

Similarly,

```
> combine(cos(x)^3);
```

$$\frac{1}{4}\cos(3\,x)+\frac{3}{4}\cos(x)$$

shows that

$$\cos^3 x = \tfrac{1}{4}\cos(3x)+\tfrac{3}{4}\cos x.$$

Finally, typing

```
> combine(cosh(x)^4);
```

$$\frac{1}{8}\cosh(4\,x)+\frac{1}{2}\cosh(2\,x)+\frac{3}{8}$$

shows that

$$\cosh^4 x = \tfrac{1}{8}\cosh(4x)+\tfrac{1}{2}\cosh(2x)+\tfrac{3}{8}.$$

Solution 1.51

(a) The commands

```
> Sp := sum(k^p,k=1..n):
> S2 := factor(eval(Sp,p=2));
```

$$S2 := \frac{n\,(n+1)\,(2\,n+1)}{6}$$

find S_2, and the other S_p may be found in a similar manner just by changing the value of `p`.

(b) One way of approaching this exercise is to use the `simplify(expr1-expr2)` construction outlined on page 42. In each case the output is 0, so that the stated relationship is verified.

```
> simplify(S3-S1^2);
```
$$0$$

```
> simplify(S5-S1^2/3*(4*S1-1));
```
$$0$$

```
> simplify(S7-S1^2/3*(6*S1^2-4*S1+1));
```
$$0$$

Solution 1.52

To derive the result for S_7, proceed as follows:

```
> Sp := sum(k^p,k=1..n):
> S1 := factor(eval(Sp,p=1)):
> S7 := factor(eval(Sp,p=7)):
> S7 := simplify(S7,{S1=a1});
```

$$S7 := 2\,a1^4 - \frac{4}{3}\,a1^3 + \frac{1}{3}\,a1^2$$

```
> factor(%);
```

$$\frac{a1^2\,(6\,a1^2 - 4\,a1 + 1)}{3}$$

The results for S_9 and S_{11} follow in exactly the same way.

Solution 1.53

(a) Following the method in the text, type

```
> Sp := sum(k^p,k=1..n):
> S1 := factor(eval(Sp,p=1)):
> S2 := factor(eval(Sp,p=2)):
> S6 := factor(eval(Sp,p=6)):
> S6 := factor(simplify(S6,{S1=a1,S2=a2}));
```

$$S6 := \frac{a2\,(-6\,a1 + 1 + 12\,a1^2)}{7}$$

The required result for S_6 is obtained.

(b) The result for S_8 follows in a similar manner.

Solution 1.54

First assign the side relations to `sd` by typing

```
> sd := {a+b+c=u,a^2+b^2+c^2=v,
         a^3+b^3+c^3=w};
```

$$sd :=$$
$$\{a+b+c=u,\ a^2+b^2+c^2=v,\ a^3+b^3+c^3=w\}$$

and then each result follows.

(a)

```
> simplify(a^4+b^4+c^4,sd);
```

$$\frac{1}{6}\,u^4 + \frac{1}{2}\,v^2 - u^2\,v + \frac{4}{3}\,w\,u$$

(b)

```
> simplify(a^5+b^5+c^5,sd);
```

$$\frac{1}{6}\,u^5 + \frac{5}{6}\,w\,u^2 - \frac{5}{6}\,u^3\,v + \frac{5}{6}\,w\,v$$

Solution 1.55

The six analogous results are obtained as follows:

```
> simplify(cosh(x)^2+sinh(x)^2);
```

$$2\cosh(x)^2 - 1$$

```
> combine(%);
```

$$\cosh(2\,x)$$

```
> simplify(cosh(x)^2-sinh(x)^2);
```

$$1$$

```
> expand(sinh(2*x));
```

$$2\sinh(x)\cosh(x)$$

```
> expand(cosh(2*x));
```

$$2\cosh(x)^2 - 1$$

```
> expand(cosh(x+y));
```

$$\cosh(x)\cosh(y) + \sinh(x)\sinh(y)$$

```
> expand(sinh(x+y));
```

$$\sinh(x)\cosh(y) + \cosh(x)\sinh(y)$$

These demonstrate the following identities:

$$\cosh^2 x + \sinh^2 x = \cosh(2x),$$
$$\cosh^2 x - \sinh^2 x = 1,$$
$$\sinh(2x) = 2\sinh x \cosh x,$$
$$\cosh(2x) = 2\cosh^2 x - 1,$$
$$\cosh(x+y) = \cosh x \cosh y + \sinh x \sinh y,$$
$$\sinh(x+y) = \sinh x \cosh y + \sinh y \cosh x.$$

What you may have noticed is that a relationship between hyperbolic functions can be obtained from a trigonometric identity by converting cos into cosh and sin into sinh, and then changing the sign of any term that contained the product of two sin terms. This is a demonstration of what is known as Osborn's rule. This rule is proved by establishing the results $\cos(ix) = \cosh x$ and $\sin(ix) = i \sinh x$, which again can be verified using Maple – a detailed proof is outside the scope of this unit.

Solution 1.56

First define f, then differentiate it and find x_1 and x_2:

```
>   f := a*x^3/3+b*x^2+c*x+d;
```

$$f := \frac{a\,x^3}{3} + b\,x^2 + c\,x + d$$

```
>   fp := diff(f,x);
```

$$fp := a\,x^2 + 2\,b\,x + c$$

```
>   (x1,x2) := solve(fp=0,x);
```

$$x1,\,x2 := -\frac{b - \sqrt{b^2 - a\,c}}{a},\ -\frac{b + \sqrt{b^2 - a\,c}}{a}$$

Now evaluate $f(x_1) - f(x_2)$ (but the output is lengthy so we suppress it here):

```
>   eval(f,x=x1)-eval(f,x=x2):
```

The output can be simplified and assigned to LHS as follows (note that the variable `lhs` is protected and cannot be assigned to, so we use LHS instead):

```
>   LHS := simplify(%);
```

$$LHS := -\frac{4\,(b^2 - a\,c)^{(3/2)}}{3\,a^2}$$

Now assign $p(x_1 - x_2)^3$ to RHS by typing

```
>   RHS := p*(x1-x2)^3;
```

$$RHS := p\left(-\frac{b - \sqrt{b^2 - a\,c}}{a} + \frac{b + \sqrt{b^2 - a\,c}}{a}\right)^3$$

and then solve for p:

```
>   solve(LHS=RHS,p);
```

$$-\frac{a}{6}$$

Hence $p = -\frac{1}{6}a$.

Solution 1.57

(a) First, define f and then evaluate a_0, a_n and b_n, for integer n:

```
>   f := x^2;
```

$$f := x^2$$

```
>   a0 := 1/(2*Pi)*int(f,x=-Pi..Pi);
```

$$a0 := \frac{\pi^2}{3}$$

```
>   assume(n::integer):
>   an := 1/Pi*int(f*cos(n*x),x=-Pi..Pi);
```

$$an := \frac{4\,(-1)^{n\sim}}{n\sim^2}$$

```
>   bn := 1/Pi*int(f*sin(n*x),x=-Pi..Pi);
```

$$bn := 0$$

So $a_0 = \pi^2/3$, $a_n = 4(-1)^n/n^2$ and $b_n = 0$ ($n \geq 1$). The result for b_n is not surprising as x^2 is an even function.

(b) Use the expression for the Fourier series given in the question, setting $b_n = 0$ for simplicity:

```
>   FN := a0+sum(an*cos(n*x),n=1..N);
```

$$FN := \frac{\pi^2}{3} + \left(\sum_{n\sim=1}^{N}\left(\frac{4\,(-1)^{n\sim}\cos(n\sim x)}{n\sim^2}\right)\right)$$

Now evaluate this expression at $N = 2$ to find the Fourier series of order 2:

```
>   eval(FN,N=2);
```

$$\frac{\pi^2}{3} - 4\cos(x) + \cos(2\,x)$$

To find the Fourier series of order 5, type

```
>   eval(FN,N=5);
```

(We do not print the output here, as it is too lengthy.)

(c) The commands

```
>   F2 := eval(FN,N=2):
>   F5 := eval(FN,N=5):
>   F10 := eval(FN,N=10):
>   plot([f,F2,F5,F10],x=-Pi..Pi,
       legend=["f","N=2","N=5","N=10"]);
```

produce Figure 1.16 (though we have suppressed the legend to save space).

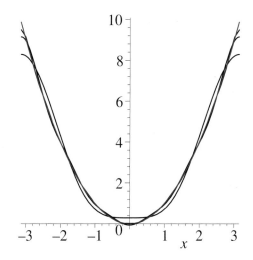

Figure 1.16

As you will be able to see if you use four colours, the higher the order of the Fourier series, the better the approximation to f.

Solution 1.58

(a) The following commands will produce Figure 1.4:

```
>   f := sin(x)/x;
```

$$f := \frac{\sin(x)}{x}$$

```
>   plot(f,x=-30..30);
```

(b) First draw the graphs of $\operatorname{sinc} x$ and $\cos x$ for $-10 \le x \le 10$ by typing

```
>  plot([f,cos(x)],x=-10..10);
```

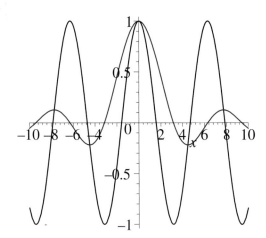

Figure 1.17

It certainly appears that the graphs meet at the points where $\operatorname{sinc} x$ is stationary. Typing

```
>  fsolve(diff(f,x)=0,x=4..5);
              4.493409458
```

shows that one of these stationary points is at $x = 4.493\,409\,458$, and this is confirmed to be where the two graphs meet with the command

```
>  fsolve(f=cos(x),x=4..5);
              4.493409458
```

which yields the same result. Similarly,

```
>  fsolve(diff(f,x)=0,x=7..8);
              7.725251837
```

and

```
>  fsolve(f=cos(x),x=7..8);
              7.725251837
```

show that there is another such point, at $x = 7.725\,251\,837$. From Figure 1.17, these are the only two such points in the interval $(0, 10]$.

(c) The given integrals evaluate to $\pi/2$:

```
>  f := sin(x)/x;
```
$$f := \frac{\sin(x)}{x}$$

```
>  int(f,x=0..infinity);
```
$$\frac{\pi}{2}$$

```
>  int(f*eval(f,x=x/3),x=0..infinity);
```
$$\frac{\pi}{2}$$

```
>  int(f*eval(f,x=x/3)*eval(f,x=x/5),
   x=0..infinity);
```
$$\frac{\pi}{2}$$

It therefore seems possible that the value of

$$\int_0^\infty \prod_{k=0}^5 \operatorname{sinc}(x/(2k+1))\,dx$$

is also $\pi/2$, and this is confirmed by typing

```
>  int(f*mul(eval(f,x=x/(2*k+1)),k=1..5),
   x=0..infinity);
```
$$\frac{\pi}{2}$$

Solution 1.59

(a) The integral is evaluated by typing

```
>  int(x^4*(1-x)^4/(1+x^2),x=0..1);
```
$$\frac{22}{7} - \pi$$

(b) From the Maple commands given in the question, we know that on $(0, 1)$,

$$\tfrac{1}{2} < \frac{1}{1+x^2} < 1.$$

Also, for $0 < x < 1$, we have $x^4 > 0$ and $(1-x)^4 > 0$. Hence

$$\tfrac{1}{2}x^4(1-x)^4 < \frac{x^4(1-x)^4}{1+x^2} < x^4(1-x)^4,$$

and the result follows.

(c) First, remove any assumptions placed on x (in the question):

```
>  x := 'x':
```

Then evaluate J and obtain the required bounds on π:

```
>  J := int(x^4*(1-x)^4,x=0..1);
```
$$J := \frac{1}{630}$$

```
>  solve({J/2=22/7-p,22/7-q=J},{p,q});
```
$$\{p = \frac{3959}{1260},\; q = \frac{1979}{630}\}$$

```
>  assign(%);
>  upper_bound := max(p,q);
```
$$upper\_bound := \frac{3959}{1260}$$

```
>  lower_bound := min(p,q);
```
$$lower\_bound := \frac{1979}{630}$$

Here we have used the commands `max` and `min` which we mentioned in introducing the exercise. An alternative way to proceed would be to omit the use of these commands and observe that J is positive, so that $\tfrac{1}{2}J < J$. Since $\tfrac{1}{2}J = \tfrac{22}{7} - p$ and $J = \tfrac{22}{7} - q$, it follows that $q < p$.

(d) To find R_5, type the following:

```
>  restart:
>  n := 5:    N := 4*n:
>  R5 := solve(int(x^N*(1-x)^N/2^(N/2-2)
   /(1+x^2),x=0..1)=(-1)^n*(Pi-Rn),Rn);
```
$$R5 := \frac{26856502742629699}{8548690331301120}$$

(e) The bounds on π can be found as follows:

```
>   J := int(x^N*(1-x)^N/2^(N/2-2),x=0..1);
```

$$J := \frac{1}{1446837166494720}$$

```
>   solve({J/2=(-1)^(N/4)*(p-R5),
    (-1)^(N/4)*(q-R5)=J},{p,q});
```

$$\{q = \frac{338805111522405359}{107845016487183360},\; p = \frac{8808932899582540303}{2803970428666767360}\}$$

```
>   assign(%);
>   upper_bound := max(p,q);
```

$$upper\_bound := \frac{8808932899582540303}{2803970428666767360}$$

```
>   lower_bound := min(p,q);
```

$$lower\_bound := \frac{338805111522405359}{107845016487183360}$$

Hence we see that

$$\frac{338\,805\,111\,522\,405\,359}{107\,845\,016\,487\,183\,360} < \pi < \frac{8\,808\,932\,899\,582\,540\,303}{2\,803\,970\,428\,666\,767\,360}.$$

Evaluating these bounds to 30 s.f., we obtain:

```
>   evalf[30](upper_bound);
```

$$3.14159265358979344527186251108$$

```
>   evalf[30](lower_bound);
```

$$3.14159265358979309969047188966$$

These bounds give an estimate for π correct to 15 decimal places.

(As in part (c), it is possible to avoid using the `max` and `min` commands.)

UNIT 2 More Maple

Study guide

Sections 2.1–2.4 each build upon the results of the previous sections and should be studied in numerical order. Sections 2.5 and 2.6 are largely independent and can be studied in either order after Section 2.4.

Access to a computer is required to study this unit.

Introduction

In *Unit 1* you met some basic Maple commands. Generally, these commands converted an expression to another expression by some process (e.g. simplifying, differentiating, integrating, etc.). In this unit we consider commands which create and manipulate objects that are more structured, such as lists, sets, vectors and matrices. This will greatly increase the range of problems that you will be able to solve in Maple.

The unit starts by looking at how functions can be defined in Maple. Maple functions are needed when an expression needs to be evaluated repeatedly – such as when manipulating lists of data. Maple functions are very closely related to Maple procedures, which you will meet in *Unit 3*.

Section 2.2 looks at some of the ways that Maple can store collections of things (data, variables, equations, etc.) and discusses some methods of manipulating them.

Section 2.3 introduces the concept of a power series and explains how Maple can truncate such a series to a polynomial.

Unit 1 introduced line graphs produced using the `plot` command. In Section 2.4 we describe some of the other types of graph that Maple can produce, such as the 3D-surface plot shown in Figure 2.1.

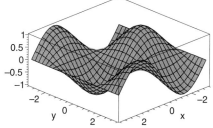

Figure 2.1 `plot3d` output

Some of the more specialized graphics commands are contained within a 'Maple package'. Maple packages are an important feature of Maple in that they contain commands that extend Maple to address problems in particular problem areas. Maple has many packages (see `?index/packages` for a list), but in this course we concentrate on the core of Maple and mention only a few packages. Subsection 2.4.3 describes how to access and use the commands that reside in packages.

The first four sections of this unit contain basic Maple commands that are used in many application areas. Section 2.5 examines some of the tools that are available in Maple for linear algebra calculations. These tools are applied to solve problems such as classifying the stationary points of surfaces like the one shown in Figure 2.1.

Section 2.6 explores how Maple can be used to solve differential equations. There are many applications of differential equations. In this unit we look at an application that you may not have studied before: the motion of rockets (Figure 2.2). The derivation of the fundamental equation (called the *rocket equation*) is not the purpose of this unit, but is included as an Appendix (Section 2.8). Some of the differential equations derived from applying the rocket equation can be solved analytically, and we show how Maple can find these solutions. Other differential equations derived from applying the rocket equation cannot be solved analytically, and we show how Maple can be used to find numerical solutions of such equations.

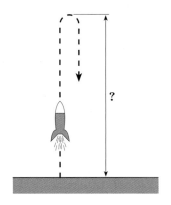

Figure 2.2   A rocket problem

2.1   Functions

The basic quantities that Maple manipulates are expressions. For example, if you want to plot a graph of the function $f(x) = x^2$, then the best way of achieving this is to plot the expression x^2 (as you did with other expressions in *Unit 1*) rather than defining a Maple function. However, there are occasions when the aim is to repeatedly evaluate an expression, and it is on these occasions that it is appropriate to define a Maple function. This section concerns the definition of Maple functions; the application of these functions to transforming data is deferred until the next section.

Maple functions are a special case of Maple procedures, which you will meet in *Unit 3*.

Defining functions

Unit 1 introduced the use of Maple built-in functions such as `sin` and `cos`. For example, to compute the sine of $\pi/3$ radians, type

```
>  sin(Pi/3);
```
$$\frac{\sqrt{3}}{2}$$

The Maple function `sin` is a function of one variable: it takes one **argument** and returns an expression called the **result** (in this case the result is $\sqrt{3}/2$). To define a function `f` that will behave exactly the same as the built-in function `sin`, you can type

```
>  f := x->sin(x);
```
$$f := x \to \sin(x)$$

The above statement is an assignment statement. On the left-hand side of the assignment is the name of the function. On the right-hand side is the **arrow notation** ->, which reads as 'maps to'. So the above Maple reads as 'the function "x maps to the sine of x" is assigned to f'. We can check this definition by applying the function to an argument, e.g.

The arrow on the right-hand side of this assignment is obtained by typing a hyphen followed by a 'greater than' sign. For more information, type ?->.

```
>  f(Pi/3);
```
$$\frac{\sqrt{3}}{2}$$

The Maple function `f` has produced the same output as `sin`. Here the function `f` could have been defined by `f:=sin` because the expression on the right-hand side of the arrow notation is so simple, but this cannot be done for the function in the following exercise.

Exercise 2.1

Define the function

$$f(x) = \tfrac{1}{2}(x-1)^2$$

as a Maple function, and evaluate this function at -1, 0 and 1.

Warning

A common mistake in Maple is to attempt to define a function f by `f(x) := x^2` rather than using `f := x->x^2`. If you subsequently type `f(x)`, Maple will duly reply with x^2, but if you type `f(y)` or `f(1)`, it will be returned 'unevaluated'. In other words, Maple has accepted a value for the string of symbols '`f(x)`' rather than a definition of a function f – using `f(x) := x^2` does *not* define a function f.

Functions of two arguments, x and y, are defined in Maple as follows:

```
>   f := (x,y)->x^2+y^2;
```

$$f := (x,\, y) \rightarrow x^2 + y^2$$

The round brackets around `x,y` are essential.

This can be used in the standard mathematical way; for example,

```
>   f(1,2);
```

$$5$$

Now try to define functions of two variables yourself.

Exercise 2.2

Define the following as functions of two variables.

(a) $f(x,y) = x^y$ (b) $z = \sin x \cos y$ (c) $u^3 + 2uv + v^2$

Check your answers by evaluating your functions at a few suitable points.

Composition of functions

The function $f(x) = \sin(\ln x)$ is a composite of the two functions $x \longmapsto \ln x$ and $x \longmapsto \sin x$. This function can be defined in Maple as a composite function using the **composition operator** `@`, as follows:

```
>   f := sin@ln;
```

$$f := \sin@\ln$$

The notation `f @ g` is the Maple analogue of the standard mathematical notation $f \circ g$.

Check:

```
>   f(2);
```

$$\sin(\ln(2))$$

Use the composition operator in the following exercise.

Exercise 2.3

Define the following functions as composite functions.

(a) $\sin(x^2 + 1)$ (b) $y^2 + 1$ where $y = \cos x$

(c) $(\ln(u))^2$ (d) $\ln(u^2)$

Composing a function with itself leads to some interesting phenomena. Maple has an operator for repeatedly composing a function with itself, namely @@. For example, the Maple expression (f@@3)(x) is equivalent to f(f(f(x))).

Some of these phenomena are described in Block B of this course.

As an example, consider the result of pressing the cosine button repeatedly on an older type of calculator. By experimenting with this you can convince yourself that, whatever number you start with, the result of pressing cosine many times is a constant (the value of the constant depends on whether you are working in degrees or radians – Maple uses radians). To calculate this constant in Maple, you can type the following:

You may need to enter this as cos(Ans) on an algebraic notation calculator.

```
>   (cos@@100)(0.0);
                    0.7390851332
```

The round brackets in this expression are essential.

This shows the result of the equivalent of pressing the cosine button one hundred times with 0.0 as the starting value. Try using the repeated composition operator to solve the following problem.

Exercise 2.4

Try iterating the function $x \longmapsto x/2 + 1/x$ twenty times, starting with different numerical values. What is the result?

Differentiation of functions

Maple provides an operator D in order to differentiate a function (as opposed to using diff to differentiate an expression). For example, consider the following Maple that expresses the well-known fact that the derived function of the sine function is cosine:

The Maple expression D(f) is more commonly written in mathematical notation as f'.

```
>   D(sin);
                    cos
```

Note that sin is the function whereas sin(x) is an expression that is obtained by applying the function sin to the argument x. The function sin is differentiated using D(sin), whereas the expression sin(x) is differentiated using diff(sin(x),x). Suppose now that we want the value of this derivative evaluated at zero: using the D operator, this can be achieved by the expression D(sin)(0), whereas using diff this would be eval(diff(sin(x),x),x=0).

The notation D(y)(0)=0 representing $y'(0) = 0$ will be used later when specifying initial conditions for differential equations.

Exercise 2.5

Define the following functions and use the `D` operator to differentiate them.

(a) $y = \sin(\cos(x))$ (b) $z = (1+t)^3$

Converting expressions to functions

The arrow notation is very convenient, but it does not suffice in all situations: the Maple command `unapply` is used in those situations. One such situation is when defining a function to be the result of a computation. For example, consider the following (simple) algebraic manipulation:

> The name `unapply` is meant to suggest that it is the opposite of `apply`, which is rarely used since `apply(f,x)` is equivalent to `f(x)`.

```
>  sol := solve(a*x+b=y,x);
```

$$sol := -\frac{b-y}{a}$$

Now define a function f that maps y to the above expression:

```
>  f := unapply(sol,y);
```

$$f := y \rightarrow -\frac{b-y}{a}$$

Values of the solution for different values of y can be computed using f:

```
>  f(1);
```

$$-\frac{b-1}{a}$$

To illustrate the necessity of using `unapply` here, consider what happens when the arrow notation is used to define a function as follows:

```
>  g := y->sol;
```

$$g := y \rightarrow sol$$

Compare the output from this function definition to the output using `unapply` above. You will see that `unapply` has replaced `sol` with its value, whilst the arrow notation has not. The function g is a constant function that maps every input `y` into the (constant) expression `sol`; for example:

```
>  g(1);
```

$$-\frac{b-y}{a}$$

The following output shows that `g` evaluates to the current value of `sol`, whereas `f` computes the required solution:

```
>  sol := 0:  g(1), f(1);
```

$$0, -\frac{b-1}{a}$$

Exercise 2.6

Use Maple to find the function h such that $h(y)$ is the solution (in x) of the equation $\sinh(2x) - 3 = y$, and evaluate $h(0)$.

2.2 Expression sequences, lists and sets

Unit 1 introduced the most basic Maple data types: fractions, floating-point numbers and expressions. In this section we introduce in successive subsections three more data types: expression sequences, lists and sets.

2.2.1 Expression sequences

Expression sequences are expressions separated by commas, for example

```
>  1,x+y,z;
```
$$1, x + y, z$$

The three expressions in the above sequence seem unrelated, but more often the expressions have a simple underlying pattern, such as the sequence $1, 4, 9, 16, 25, 36, 49, 64, 81, 100$, consisting of the first ten square numbers. Maple provides `seq` for generating such sequences:

```
>  seq(i^2,i=1..10);
```
$$1, 4, 9, 16, 25, 36, 49, 64, 81, 100$$

seq is a shortening of sequence and is usually pronounced 'seek'.

Here `seq` has two arguments. The first argument is an expression, and the second argument specifies the variable which is changing on the left-hand side of the '=', and the range of variation on the right-hand side. The following exercise gives you some practice in using `seq`.

Exercise 2.7

Use `seq` to generate the following expression sequences.

(a) $2, 4, 6, 8, 10$

(b) $1, 3, 5, 7, 9, 11, 13, 15, 17, 19$

(c) $2, 3, 5, 7, 11, 13, 17, 19, 23, 29$ [*Hint*: Use the function `ithprime`.]

(d) $1, x, x^2, x^3, x^4, x^5$

(e) $\sin(n\pi/2), \quad n = 0, 1, 2, \ldots, 9$

Consider the expression sequence v defined as follows:

```
>  v := seq(2^i,i=0..10);
```
$$v := 1, 2, 4, 8, 16, 32, 64, 128, 256, 512, 1024$$

To extract the fifth element of v we use **indexing**: the position of the expression we want to extract (in this case 5 for the fifth in the list) is called the **index**. It is enclosed in square brackets after the expression name, as follows:

Indexing is used in exactly the same way to extract elements of lists, sets, vectors and matrices.

```
>  v[5];
```
$$16$$

The result of a function can be an expression sequence. The following function maps the xy-plane to itself, and represents a translation of one unit along the x-axis:

```
>  f := (x,y)->(x+1,y);
```
$$f := (x, y) \rightarrow (x + 1, y)$$

All the brackets here are essential: the assignment `f := (x,y)->x+1,y` makes `f` an expression sequence (first element the function `(x,y)->x+1`, second element `y`).

72

This function is used as follows:

```
>  f(1,2);
```
$$2, 2$$

A *Fibonacci sequence* is obtained by repeatedly adding two consecutive elements to obtain the next element.

(a) Define a function `fs` of two arguments `a` and `b` that returns `b` and `a+b` (so that if a and b are two consecutive elements of a Fibonacci sequence, then so are b and $a + b$).

(b) Use the repeated composition operator `@@` to compute the 100th element in the Fibonacci sequence that starts $1, 1, 2, 3, 5, 8, 13, \ldots$.

The Fibonacci sequence that starts with $1, 1$ is often referred to as *the* Fibonacci sequence.

A useful feature of `seq` is that we can step through a sequence in steps of size other than 1. Thus, for example, the odd square numbers up to 361 can be generated as follows:

```
>  seq(i^2,i=1..19,2);
```
$$1, 9, 25, 49, 81, 121, 169, 225, 289, 361$$

The final '2' in the input indicates that we step through the sequence $1, \ldots, 19$ in steps of size 2.

Expression sequences are very useful, but they do have drawbacks. In particular, the Maple manual warns that 'Passing expression sequences into functions can lead to unexpected results', as shown in the next subsection. Now we move on to look at a data type that can be passed safely to functions.

2.2.2 Lists

In Maple, a **list** is distinguished from an expression sequence by being enclosed in square brackets: for example, `[1,2,3]` is a list. The following statement creates a list from an expression sequence by enclosing the latter in square brackets:

```
>  L := [seq(i^2,i=1..10)];
```
$$L := [1, 4, 9, 16, 25, 36, 49, 64, 81, 100]$$

The reverse operation of creating an expression sequence from a list is achieved by using `op`:

```
>  op(L);
```
$$1, 4, 9, 16, 25, 36, 49, 64, 81, 100$$

The `op` command has many uses, but here we are highlighting only one of them.

A related command `nops` counts the number of elements in a list:

```
>  nops(L);
```
$$10$$

`nops` is short for 'number of operands'.

A fundamental difference between expression sequences and lists can by exemplified using `nops`. Consider again the expression sequence

```
>  v := seq(2^i,i=0..10):
```

The command `nops(v);` will produce an error message, whereas `nops([v]);` will produce the expected answer 11.

Lists can be indexed (e.g. `L[2]`) in the same way as expression sequences.

Exercise 2.9

Define a function called **reverse** with a single list L as an argument. The result of the function should be a list of the elements of L in reverse order. Test your answer by reversing the list $[1, 2, 3, 4, 5]$ to obtain $[5, 4, 3, 2, 1]$.

[*Hints*: Use **seq** to construct the list by indexing. The index of the last element of the list is given by **nops(L)**. Note that when using **seq**, the step size can be negative.]

Exercise 2.10

Define a function **join** that takes two lists as arguments and returns the list that results from amalgamating the two lists; for example:

```
>  join([1,2],[3,4]);
```
$$[1,\ 2,\ 3,\ 4]$$

Lists can be manipulated by using **select** to extract elements from a given list. For example, consider the problem of creating a list of the prime numbers less than twenty. The following uses the function **isprime** as a selection criterion:

```
>  select(isprime,[seq(i,i=1..20)]);
```
$$[2,\ 3,\ 5,\ 7,\ 11,\ 13,\ 17,\ 19]$$

isprime is a built-in function that returns true *if and only if the argument is prime, e.g.* isprime(3) *returns* true *and* isprime(4) *returns* false.

The output of **select** is a list of those integers which are prime, i.e. the function **isprime** is applied to each element of the list in turn, and the result is a list of the elements for which **isprime** returns *true*.

The above example uses the built-in function **isprime**, but **select** is more useful with user-defined functions such as the following:

```
>  gt20 := x->x>20;
```
$$gt20 := x \rightarrow 20 < x$$

This function registers *true* if the argument is greater than 20; for example, gt20(21) registers *true*, while gt20(19) registers *false*. Using this function, we can construct a list of those of the first fifteen prime numbers that are greater than 20. First create a list of the first fifteen prime numbers:

```
>  L15 := [seq(ithprime(i),i=1..15)];
```
$$L15 := [2,\ 3,\ 5,\ 7,\ 11,\ 13,\ 17,\ 19,\ 23,\ 29,\ 31,\ 37,\ 41,\ 43,\ 47]$$

To force Maple to evaluate gt20(19), *type* evalb(gt20(19)). *The b is for 'boolean', since a* boolean variable *is one that takes the values true, false.*

Now use **select** to choose the elements greater than 20:

```
>  select(gt20,L15);
```
$$[23,\ 29,\ 31,\ 37,\ 41,\ 43,\ 47]$$

This is the required list of prime numbers.

Expressions returning *true* or *false* can be constructed using expressions built from the six **comparison operators** listed in Table 2.1. Use **select** with user-defined functions to do the following exercises.

Table 2.1
Comparison operators

<	less than
<=	less than or equal to
>	greater than
>=	greater than or equal to
=	equal to
<>	not equal to

Exercise 2.11

Use `select` to create the list of:

(a) those prime numbers that are between 100 and 200;

(b) those roots of the equation $x^4 + x^2 + 1 = 0$ that have positive real part;

(c) the solution of the simultaneous equations $x + 3y = 1$ and $x^2 + y^2 = 1$ with y non-zero.

Exercise 2.12

Twin primes are pairs of prime numbers that have difference two, such as 3 and 5, or 11 and 13. This exercise is to construct a list of twin primes.

(a) Use the Maple function `ithprime` to construct a list of the first one hundred prime numbers.

(b) Use `select` to filter the list produced in part (a) to produce a new list consisting of the smaller member of each twin-prime pair.

[*Hint*: Use the function `isprime` in your selection criterion.]

A list can be transformed by applying a function to every element of the list using `map`. For example, the following Maple statements create a list L and then create a list consisting of the square root of each element of L:

```
>  L := [seq(i,i=4..16)]:  map(sqrt,L);
```
$$[2, \sqrt{5}, \sqrt{6}, \sqrt{7}, 2\sqrt{2}, 3, \sqrt{10}, \sqrt{11}, 2\sqrt{3}, \sqrt{13}, \sqrt{14}, \sqrt{15}, 4]$$

Use `map` in the following exercise.

Exercise 2.13

Use `map` to create:

(a) the list $[1, 3, 5, 7, 9]$ from the list $[0, 1, 2, 3, 4]$;

(b) the list $[1, x, x^2, x^3, x^4]$ from the list $[0, 1, 2, 3, 4]$.

2.2.3   Sets

Sets are unordered sequences without repetition. Maple uses the standard mathematical notation: sets are enclosed in curly brackets. An expression sequence is converted into a set by enclosing it in curly brackets, for example:

```
>  S := {4,4,4,4,3,3,3,2,2,1};
```
$$S := \{1, 2, 3, 4\}$$

The repetition in the expression sequence is removed in the output, i.e. each number appears only once. The order of the input elements is not, in general, preserved in the output. The order that appears is the order of the Maple internal representation of the elements, which is so complicated that it can appear random (although small integers usually appear in numerical order).

All of the operations that can be applied to lists can also be applied to sets. For example, `op(S)`, `nops(S)`, `select(isprime,S)` and `map(sqrt,S)` work as expected (the latter two will produce sets as output rather than lists).

Try the following exercise using sets.

Exercise 2.14

An integer k is a *perfect power* if there exist integers $m \geq 2$, $n \geq 2$ such that $k = m^n$.

(a) Calculate the set P of perfect powers for $m \leq 1000$ and $n \leq 10$.

(b) Compute the sum

$$\sum_{k \in P} \frac{1}{k-1},$$

and compare it with the sum

$$\sum_{m=2}^{1000} \sum_{n=2}^{10} \frac{1}{m^n - 1},$$

accounting for any differences.

2.3   Series

A useful mathematical technique is the representation of a function in terms of a **power series**, i.e. as a sum of powers of a variable. Power series are a generalization of the Taylor series you have already met.

Consider the Taylor series of $\ln(1 + x)$, expanded about $x = 0$:

$$\ln(1 + x) = x - \frac{1}{2}x^2 + \frac{1}{3}x^3 + \cdots + \frac{(-1)^{n-1}}{n} x^n + \cdots.$$

In Maple a truncated form of this is calculated using the `series` command. This command has three arguments: the function being expanded (here $\ln(1 + x)$), the expansion point (here $x = 0$), and the order at which the series is truncated (5 in the following example):

The `series` command returns the Taylor series if it exists, but otherwise may return a more general series.

```
>  lns := series(ln(1+x),x=0,5);
```

$$lns := x - \frac{1}{2}x^2 + \frac{1}{3}x^3 - \frac{1}{4}x^4 + O(x^5)$$

Note the presence of the **order term**, $O(x^5)$, in the Maple output. This signifies that terms with powers equal to and higher than x^5 have been neglected.

We can expand about other points by changing the second argument; thus the expansion of $\sin x$ about $x = \pi/2$ is given by

```
>   sns := series(sin(x),x=Pi/2,5);
```

$$sns := 1 - \frac{1}{2}\left(x - \frac{\pi}{2}\right)^2 + \frac{1}{24}\left(x - \frac{\pi}{2}\right)^4 + O\left(\left(x - \frac{\pi}{2}\right)^6\right)$$

This series has only even terms, so the order term is $O\left((x - \frac{\pi}{2})^6\right)$ rather than $O\left((x - \frac{\pi}{2})^5\right)$.

and since $\sin(\pi/2 + x) = \cos x$, we see that this has returned the Taylor series expansion of $\cos z$, i.e. $1 - z^2/2! + z^4/4! + O(z^6)$, where $z = x - \pi/2$.

The order of the expansion is changed simply by altering the last argument: thus the expansion of $\ln(1 + x)$ which includes the term in x^7 is given by

```
>   lns := series(ln(1+x),x=0,8);
```

$$lns := x - \frac{1}{2}x^2 + \frac{1}{3}x^3 - \frac{1}{4}x^4 + \frac{1}{5}x^5 - \frac{1}{6}x^6 + \frac{1}{7}x^7 + O(x^8)$$

To extract the coefficients of a series, use `coeff` in the same way as `coeff` was used with polynomials in *Unit 1*. For example, the coefficient of x^5 is given by

```
>   coeff(lns,x,5);
```

$$\frac{1}{5}$$

The order term is useful for some operations on series (e.g. differentiating and integrating) as it keeps track of the truncation order of the series. However, for other operations (e.g. plotting) it is necessary to convert the series to a polynomial as follows:

```
>   lnp := convert(lns,polynom);
```

$$lnp := x - \frac{1}{2}x^2 + \frac{1}{3}x^3 - \frac{1}{4}x^4 + \frac{1}{5}x^5 - \frac{1}{6}x^6 + \frac{1}{7}x^7$$

Exercise 2.15

In all the following, the expansion point is $x = 0$.

(a) Find the series expansion of $\ln(1 + x - \sin x)$ up to and including the term in x^7.

(b) Find the series expansion of $2\sinh x \cosh x$ and $\sinh(2x)$ up to and including the term in x^7. Subtract these series to show that, to this order, these functions are the same. Note that for the subtraction, the series will have to be converted to polynomials.

(c) Find the series expansion of $\ln(1 + \sqrt{x} + x)$ up to and including the term in x^3.

(d) Find the series expansion of $\cot x$ up to and including the term in x^3.

2.4 Graphics

Unit 1 introduced `plot` to draw the graph of an expression. This section considers more of the possibilities for creating graphics with Maple.

2.4.1 Plotting functions

You saw in *Unit 1* that it is possible to plot a function by assigning a variable to an expression; for example, the graph of part of a parabola will be obtained by using the commands `y:=x^2: plot(y,x=-1..1);`.

It is possible to do much the same by defining the Maple *function* `f:=x->x^2`. Then either of the commands `plot(f(x),x=-1..1);` or `plot(f,-1..1);` will give the output in Figure 2.3 (the first with an x label on the x-axis and the second without).

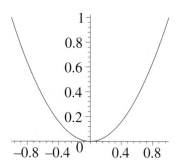

Figure 2.3   Plot of $y = x^2$

The axes on your screen will not be marked exactly as in Figure 2.3; here and elsewhere we have adjusted the plot for readability.

However, these commands do not always give the same result; sometimes the first command fails. Usually this is because the Maple function `f(x)` can be evaluated only if `x` is a number.

To take a simple example, suppose that $f(x)$ is the sum of the positive integers that are less than or equal to the (real) number x:

$$f(x) = \sum_{k=1}^{[x]} k.$$

The notation $[x]$ means 'the largest integer less than or equal to x'.

One Maple representation of this function is

```
>   f := x->add(k,k=1..x);
```
$$f := x \rightarrow \mathrm{add}(k,\, k = 1 \mathinner{.\,.} x)$$

This is what happens if we attempt to plot its graph using the `plot` command as in *Unit 1*:

```
>   plot(f(x),x=1..10);
```
`Error, (in f) unable to execute add`

The reason that this command fails is that, inside `plot`, Maple tries to evaluate `f(x)` to an expression, subsequently evaluating this expression numerically at the sample points. This causes problems because `add` works only when the range is specified numerically. This difficulty can be avoided by using the version of `plot` that takes `f` as a function rather than `f(x)` as an expression:

```
>   plot(f,1..10);
```

The output is shown in Figure 2.4.

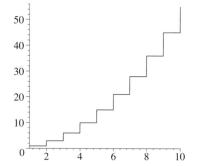

Figure 2.4   Plotting sums of integers

An alternative device is to place `f(x)` in quotes to delay the evaluation of `f(x)` until `x` has a numeric value. For example, `plot('f(x)',x=1..10);` will also produce the output shown in Figure 2.4 (but with the x-axis labelled).

Note that both quotes are 'end of quote' marks.

Exercise 2.16

For the function

$$f(x) = \prod_{k=1}^{[x]} k^{-1/4},$$

use the `mul` command to create a suitable Maple function, and plot its graph for $1 \le x \le 10$.

2.4.2 Plotting coordinates

The Cartesian coordinates of a point can be represented in Maple as a list. So the point $(1, 2)$ in the xy-plane can be represented by the Maple list `[1,2]`. Points in three-dimensional space can be represented by lists with three elements.

The elements of lists can be any Maple data type – including lists. For example, the list `[[1,1],[2,4],[3,9]]` is a valid Maple list. This could be used to represent the list consisting of the three pairs of coordinates $(1, 1)$, $(2, 4)$ and $(3, 9)$. The above list can be created in Maple as follows:

```
> L := [seq([x,x^2],x=1..3)];
```
$$L := [[1, 1], [2, 4], [3, 9]]$$

The list L can be indexed in the usual way to obtain the individual points. For example, the second element of the list can be extracted as follows:

```
> L[2];
```
$$[2, 4]$$

The result is a list that represents the point $(2, 4)$. This resulting list can then be indexed to access either the x-coordinate or the y-coordinate. The x-coordinate of the second element of L can be extracted by using two indexing operations as follows:

```
> L[2][1];
```
$$2$$

Maple provides a shorthand notation for this double indexing operation by defining `L[i,j]` to be the same as `L[i][j]`.

Data points representing points in the plane can be plotted using `plot`. For example, the points $(1, 1)$, $(1, 2)$, $(2, 2)$, $(2, 1)$ form the corners of a square, which can be plotted as follows (notice how the first point is repeated in the list in order to 'close' the square, and how the axis ranges have been specified using extra parameters):

```
> plot([[1,1],[1,2],[2,2],[2,1],[1,1]],0..3,0..3);
```

The output is shown in Figure 2.5.

Note that in this plot, successive pairs of plot points are joined by lines. In the following subsection you will see how to plot points without joining them in this way.

Now try the following exercise.

Maple also provides a triple indexing notation `L[i][j][k]` or `L[i,j,k]` to access individual elements of lists of lists of lists.

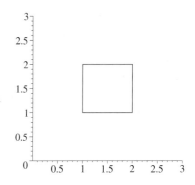

Figure 2.5 Plot of a square

Exercise 2.17

Create a list of the coordinates of the vertices of a regular pentagon that can be inscribed in a unit circle. Use your list to plot a regular pentagon.

2.4.3 Maple packages

In order to save memory, a wide variety of specialized Maple commands are grouped together so that the user can load them only when necessary. There are many such **packages**, and each can be loaded as required using the `with` command. In this subsection we concentrate on the `plots` package, which allows a wide variety of graphs to be created. A full, and rather daunting, list of all the packages can be viewed by typing `?index/packages` at the prompt. All packages are invoked using the same type of command, and if you wish to use the command of any package it is as well to invoke it immediately after the `restart` at the head of the worksheet; however, a package can be invoked at any point in a worksheet.

The plots package is invoked as follows:

```
>   restart:
>   with(plots):
```

```
Warning, the name changecoords has been redefined
```

Notice that a silent terminator has been used: without this, Maple outputs a long list of the commands now available. (We shall use just three of these here.)

Depending on your version of Maple, you may see this warning message about `changecoords`, which can be ignored.

The `display` command is used to combine the plots created using other plot commands. One use is when plotting a line graph with the graph marked at specific values of the independent variable: for instance, if $y = \sin(6\pi \exp(-(x - \pi)^2))$ and we wish to plot the graph of y for $0 < x < 2\pi$ and to mark the points $y(n)$, $n = 1, 2, \ldots, 6$, with a square, we proceed as follows. First invoke `with(plots)`:

```
>   restart:  with(plots):
```

Now define the function:

```
>   y := x->sin(6*Pi*exp(-(x-Pi)^2)):
```

The list of points is

```
>   Ly := [seq([n,evalf(y(n))],n=1..6)]:
```

The graphical representation of these points is put in a plot structure

```
>   pL := plot(Ly,style=point,symbol=box):    ; colour = black.
```

and the ordinary graph is put in another plot structure

```
>   p := plot(y,0..2*Pi):
```

Now combine these using `display`:

```
>   display(p,pL);
```

The output is shown in Figure 2.6.

Any procedure in any package can be used without first using the appropriate `with` command, by using its long name (that is, the name of the package followed by the name of the procedure in square brackets). Thus the previous command would be replaced by

```
>   plots[display](p,pL);
```

In this case Figure 2.6 could have been produced directly with the `plot` command, since both `p` and `pL` were produced by `plot`, by using:

```
>   plot([y(x),Ly],x=0..2*Pi,style=[line,point],
    symbol=[box,box]);
```

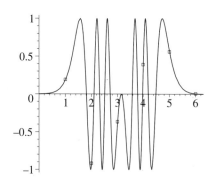

Figure 2.6 Combined graphs plotted using `display`

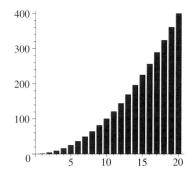

The `Statistics` package includes graph types that are commonly used in statistics (histograms, pie charts, etc. – type `?Statistics/Visualization` for a list). For example, the following produces a bar chart from a list:

```
> Statistics[ColumnGraph]([seq(n^2,n=1..20)],colour=red);
```

The bar chart is shown in Figure 2.7.

Figure 2.7   `ColumnGraph` output

2.4.4 Contour plots

Plots of functions of two variables can be made using either the `plot3d` command (which does *not* require `with(plots)` to be invoked) or the command `contourplot` from the `plots` package. The two outputs look quite different. For example, consider the following:

```
> with(plots):
> contourplot(-sin(x)*cos(y),x=-Pi..Pi,y=-Pi..Pi);
```

The output is shown in Figure 2.8. In this plot, heights are represented not by artistic perspective, but by contours. For comparison, suppose we use `plot3d` instead:

```
> plot3d(-sin(x)*cos(y),x=-Pi..Pi,y=-Pi..Pi,axes=boxed);
```

The output is shown in Figure 2.9.

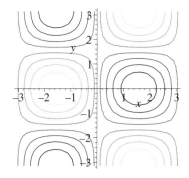

Figure 2.8   `contourplot` output

The viewpoint for the three-dimensional plot can be adjusted by clicking and dragging on the graph; it gives a good overall impression of the shape of the function, whereas the contour plot is usually preferable if more accurate information is needed.

Exercise 2.18

Plot the function
$$f(x) = x\left(x^2 - 3y^2\right)e^{1-x^2-y^2},$$
using `plot3d` or `contourplot` to locate the coordinates of the stationary points approximately. Use `fsolve` to locate the points exactly, and hence show that the stationary points (other than the one at the origin) all lie on a circle centred at the origin. Use Maple to find the radius of the circle on which these stationary points lie.

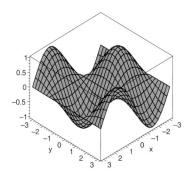

Figure 2.9   `plot3d` output

2.4.5 Implicit plots

Consider the equation $\cos(xy) = x/y$ that relates two variables x and y. It is not possible to rearrange this equation to make x or y the subject. So it is not possible to derive an expression to plot that will give an impression of how y depends on x. In this sort of situation we use `implicitplot` to visualize the solutions of an equation:

```
> with(plots):
> implicitplot(cos(x*y)=x/y,x=-1..1,y=-2..2);
```
numpoints = 1000000

The output shown in Figure 2.10 shows the relationship between x and y. It shows that for some x there are at least two values of y that satisfy the equation. This means that the mapping from x to y does not define a function, and that attempts to plot the complete relationship by first constructing a function (perhaps using `fsolve`) are certain to fail.

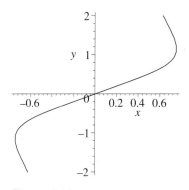

Figure 2.10   `implicitplot` output

Or use `> implicitplot (y * cos (x*y) = x, x = -1..1, y = -2..2);`

Nevertheless, in the vicinity of $x = 0$ and $y = 0$, the equation

$$\cos(xy) = \frac{x}{y}$$

does define y as a function of x. It can be shown that this dependence can be represented by the series

$$y = x + \frac{1}{2}x^5 + \frac{17}{24}x^9 + \frac{961}{720}x^{13} + \cdots \quad \text{provided that } |x| < 0.749,$$

and it is sometimes useful to compare, graphically, the first four terms of this series with the graph produced by `implicitplot`. This is achieved by producing the two graphs as plot structures, and combining these using `display` as follows.

```
> restart:  with(plots):
```

The series is

```
> ys := x+x^5/2+17*x^9/24+961*x^13/720:
```

and the two plot structures are

```
> p1 := implicitplot(cos(x*y)=x/y,x=-1..1,y=-2..2):
> p2 := plot(ys,x=-1..1,-2..2,colour=black):
```
✓ numpoints = 1 000 000

and these are combined to give

```
> display(p1,p2);
```

The output is shown in Figure 2.11.

✗ Or define as p1 := implicitplot (y * cos(x*y) = x, x = -1..1, y = -2..2): *as previously.*

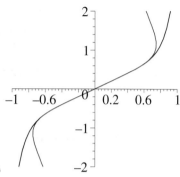

Figure 2.11 The black line depicts the series representation of $y(x)$, and the red line depicts its exact representation

Exercise 2.19

Two variables x and y are related by the equation

$$\tanh(x + y)\left(1 + 3\sin^2 y\right) = x.$$

Use `implicitplot` to picture the solutions of this equation for x and y between -4 and 4. Use your picture to estimate the largest range of x including the origin for which the mapping from x to y is single-valued.

2.4.6 Adding titles

A useful feature of Maple plot structures is the facility for adding titles. A title is just a **string** – that is, a set of symbols enclosed in double quotation marks – which is included in the `plot` command as an optional argument. Thus to plot the graph of $\sin x$ and add the title *The sine function*, we use the following command:

```
> plot(sin,-Pi..Pi,title="The sine function");
```

The output is shown in Figure 2.12.

A title can be made to change according to one or more parameter values. Suppose that we require a graphical comparison of the Nth-order Taylor series of $\sin x$ with the exact function, and wish to label each graph with the value of N.

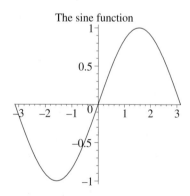

Figure 2.12 Titled `plot` output

One method of doing this is to first create a function `y(N)` representing the truncated Taylor series, using the commands introduced in Section 2.3. Note the use of `convert` so that the output is a polynomial; it is not possible to plot a series without this conversion.

```
>  y := N->convert(series(sin(x),x=0,N),polynom):
```

Thus the Taylor series up to and including the x^5 term is

```
>  y(6);
```

$$x - \frac{1}{6}\,x^3 + \frac{1}{120}\,x^5$$

Note that `y(6)` and `y(7)` give the same output, since the Taylor series for $\sin x$ has only odd terms.

In order to allow for any value of N in the graph title, we use the **concatenation operator** `||`, which concatenates strings and names: in this case we use it to concatenate the string *Truncated at N =* with the value of `N`. Thus for '$N = 4$' we could use the commands

```
>  N := 4:  plot(y(N),x=-Pi..Pi,title="Truncated at N = "||N):
```

If you are not clear how this concatenation command works, consider the statements

```
>  N := 4:  "Truncated at N = "||N;
              "Truncated at N = 4"
```

and vary the value of N.

It is sometimes more convenient to create a function of N that creates the plot structure and also compares this series with the exact function. For this we need a minor modification of the previous command:

```
>  f := N->plot([sin(x),y(N)],x=-Pi..Pi,
     title="Truncated at N = "||N):
```

For $N = 4$ the command `f(4);` gives the graph shown in Figure 2.13. A good method of understanding how this approximation behaves with N is to view an **animation**, which is produced by first creating an expression sequence of plot structures using `f(N)` and the `seq` command:

```
>  sf := seq(f(N),N=2..10,2):
>  with(plots):
>  display(sf,insequence=true);
```

The output of this command is an animation that cannot be shown here. It is viewed on the screen by placing the cursor on the graph displayed, clicking the left mouse button, which displays the Animation toolbar, then clicking on the Play button. The default speed of showing animations is quite fast; this speed can be changed using the toolbar.

If N were not an integer, then `||N` would need to be replaced by `||(convert(N,string))`; alternatively, one could use the `cat` function as described on page 183.

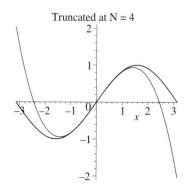

Figure 2.13 `plot` output

Step size 2 is used because each even-order Taylor polynomial for $\sin x$ is the same as the previous odd-order polynomial.

Exercise 2.20

(a) Generalize your solution to Exercise 2.17 to create a function that will plot (with a suitable title) a regular N-gon for any $N \geq 3$.

(b) Set up an animation sequence of the first twenty of these N-gons.

The concatenation operator can also be used to generate a sequence of expression names (such as coefficients of polynomials). Suppose, for example, that you wish to set up a quartic polynomial $a_0 + a_1 x + a_2 x^2 + a_3 x^3 + a_4 x^4$, the numerical values of whose coefficients are as yet unknown. This can be done as follows:

```
>  p := add(a||i * x^i,i=0..4);
```

$$p := a0 + a1\,x + a2\,x^2 + a3\,x^3 + a4\,x^4$$

83

Exercise 2.21

Use Maple to find the coefficients of the quadratic polynomial through the points $(1, 2.5)$, $(2, 1.5)$ and $(3, \pi)$, and plot the result in the range $0 \leq x \leq 4$.

2.5   Vectors and matrices

This section shows some of the linear algebra features of Maple. Many of these are contained in the **LinearAlgebra** package, invoked using the **with(LinearAlgebra)** command; but for elementary vector and matrix manipulation this is not necessary.

2.5.1 Defining vectors

Vectors are created in Maple using **Vector** (note the initial capital letter; Maple also has **vector** that is now superseded). For example, the following creates a vector v with two elements (both zero by default):

```
>  v := Vector(2);
```

$$v := \begin{bmatrix} 0 \\ 0 \end{bmatrix}$$

Maple constructs column vectors by default. To construct a row vector, use **Vector[row](2)**.

The components of v can be indexed using the usual notation; for example:

```
>  v[1] := 1:  v[2] := 4:  v;
```

$$\begin{bmatrix} 1 \\ 4 \end{bmatrix}$$

The vector v can be created from the list [1,4] as follows:

```
>  Vector([1,4]):
```

A vector can also be written by enclosing an expression sequence in angle brackets (that is, the 'less than' and 'greater than' symbols on the keyboard):

```
>  <1,4>:
```

The size parameter is optional if it can be deduced: this command is equivalent to **Vector(2,[1,4])**.

The same vector can be created using an **initializer function**, i.e. a function that defines the initial values of the elements of the vector (the function is evaluated at the index of the elements, i.e. 1 and 2 in this case):

```
>  sqr := i->i^2:  Vector(2,sqr):
```

Vectors can be initialized to have each element a given constant, as follows:

```
>  Vector(2,1);
```

$$\begin{bmatrix} 1 \\ 1 \end{bmatrix}$$

Try the following exercise to create some vectors.

Create the following:

(a) the vector $3\mathbf{i} + 2\mathbf{j} + 5\mathbf{k}$;

(b) a vector with components equal to the first five even numbers.

2.5.2 Defining matrices

The syntax of **Matrix** is very similar to that of **Vector** (note the initial capital letter; Maple also has **matrix** that is now superseded). This gives a variety of methods for creating matrices. The 2×2 zero matrix can be obtained by:

```
> Matrix(2);
```

$$\begin{bmatrix} 0 & 0 \\ 0 & 0 \end{bmatrix}$$

For square matrices only one size parameter needs to be specified, as shown here. Non-square matrices require two size parameters.

A list of lists can be used to create a matrix as follows:

```
> Matrix([[1,2,3],[4,5,6]]);
```

$$\begin{bmatrix} 1 & 2 & 3 \\ 4 & 5 & 6 \end{bmatrix}$$

In the same way as for vectors, there is a shorthand method for creating matrices which uses the angle-bracket notation. Separating elements by commas produces columns (in the same way as for vectors), and separating elements by vertical bars (|) produces rows; for example:

```
> A := <<1|2|3>,<4|5|6>>;
```

$$A := \begin{bmatrix} 1 & 2 & 3 \\ 4 & 5 & 6 \end{bmatrix}$$

The above matrix has been produced by listing the rows. It could also have been produced by listing the columns, using `<<1,4>|<2,5>|<3,6>>`.

The elements of a matrix can be accessed by using the notation introduced for lists of lists; for example, `A[2,3]` evaluates to 6.

Note that `A[2,3]` has the same meaning as the usual subscript notation A_{23}.

For the next exercise, you need to know that the Maple expression for the modulus (or absolute value) of an expression x is `abs(x)`.

Create the following:

(a) the matrix $\begin{pmatrix} 2 & 0 & 9 \\ 3 & 2 & 5 \end{pmatrix}$;

(b) a 3×3 matrix M with $M_{ij} = |i - j|$.

2.5.3 Matrix multiplication

One of the fundamental properties of matrix multiplication is that it is not commutative; that is, if **A** and **B** are two matrices that can be multiplied together in either order, then $\mathbf{AB} \neq \mathbf{BA}$ in general. Maple uses dot (.) as the symbol for non-commutative multiplication: so the product of two matrices **A** and **B** is denoted by A.B in Maple. For example, consider the following matrix multiplication:

Type ?. for more information.

```
>  A := <<1,2>|<3,4>>;
```
$$A := \begin{bmatrix} 1 & 3 \\ 2 & 4 \end{bmatrix}$$

```
>  B := <<0,1>|<1,0>>;
```
$$B := \begin{bmatrix} 0 & 1 \\ 1 & 0 \end{bmatrix}$$

```
>  A.B;
```
$$\begin{bmatrix} 3 & 1 \\ 4 & 2 \end{bmatrix}$$

The dot notation is also used for the scalar product of two vectors:

```
>  v := <1,2,3>:  v.v;
```
$$14$$

The scalar product is sometimes called the dot product, because of this common notation.

Try the following exercise, which uses the dot notation for matrix–vector multiplication.

Exercise 2.24

Let **v** be the vector $(1,0,0)^T$, and define a matrix **A** as

$$\mathbf{A} = \begin{pmatrix} 0.3 & 0.2 & 0.5 \\ 0.2 & 0.6 & 0.2 \\ 0.5 & 0.2 & 0.3 \end{pmatrix}.$$

Use the Maple function mulA := v->A.v to compute $\mathbf{A}^{100}\mathbf{v}$. What is the effect of multiplying by **A** again? Interpret this result in terms of the eigenvalues and eigenvectors of **A**.

2.5.4 Eigenvalues and eigenvectors

In this subsection we introduce features available in the LinearAlgebra package. We shall assume throughout this subsection and Subsection 2.5.5 that with(LinearAlgebra): has been invoked. Note that this assumption carries forward also to the solutions to Exercises 2.25 and 2.26.

Consider the matrix given in Exercise 2.24 above. Its eigenvectors and eigenvalues may be computed as follows (the number of digits shown has been deliberately reduced for presentation purposes):

```
>  with(LinearAlgebra):
>  A := <<0.3,0.2,0.5>|<0.2,0.6,0.2>|<0.5,0.2,0.3>>:
>  interface(displayprecision=1):
>  ev := Eigenvectors(A);
```
$$ev := \begin{bmatrix} 1.0 + 0.0\,I \\ -0.2 + 0.0\,I \\ 0.4 + 0.0\,I \end{bmatrix}, \begin{bmatrix} -0.6 + 0.0\,I & -0.7 + 0.0\,I & 0.4 + 0.0\,I \\ -0.6 + 0.0\,I & 0.4\,10^{-15} + 0.0\,I & -0.8 + 0.0\,I \\ -0.6 + 0.0\,I & 0.7 + 0.0\,I & 0.4 + 0.0\,I \end{bmatrix}$$

Notice how Maple is displaying the number 0.4×10^{-15}.

The result given by `Eigenvectors` is an expression sequence. As you may observe, `Eigenvectors` returns not only the eigenvectors but also the eigenvalues of the matrix. The arrangement is as follows: the nth element of the column vector is the eigenvalue corresponding to the eigenvector that is the nth column of the output matrix. ~~The eigenvectors are normalized so that each is a unit vector.~~ *not necessarily true.*

You may have been surprised to see that all the eigenvalues and eigenvectors are displayed as complex numbers: this is because it is the general case. In this particular case all the eigenvalues and eigenvectors are real, as is always the case for real symmetric matrices. The eigenvalues can be obtained without the spurious imaginary parts by `map(Re,ev[1]);`, which uses `map` to apply `Re` to each element of the column vector `ev[1]` of eigenvalues. Similarly, to obtain the matrix of eigenvectors without the imaginary parts, type `map(Re,ev[2]);`.

> You should remember from *Unit 1* that `Re(z)` extracts the real part from a complex number `z`.

If Maple is instructed that **A** is symmetric, then it will give real eigenvalues and eigenvectors. To enter a symmetric matrix, you should use the `shape` keyword. This allows you create the matrix by giving just the entries on and below the leading diagonal. For example, for the matrix above, we type a list consisting of lists of lengths 1, 2 and 3:

> Type `?shape` to see a list of allowed shapes.

```
>  A := Matrix([[0.3],[0.2,0.6],[0.5,0.2,0.3]],
        shape=symmetric):
```

```
>  lambda,U := Eigenvectors(A);
```

$$\lambda, U := \begin{bmatrix} -0.2 \\ 0.4 \\ 1.0 \end{bmatrix}, \begin{bmatrix} -0.7 & 0.4 & 0.6 \\ -0.1 \ 10^{-15} & -0.8 & 0.6 \\ 0.7 & 0.4 & 0.6 \end{bmatrix}$$

> Note that the order of the eigenvalues is not fixed and that Maple has returned the eigenvalues in a different order to previously. The order of the eigenvalues is still consistent with the (new) order of the eigenvectors.

Note how we have assigned the expression sequence returned by `Eigenvectors` to a sequence of names (this is permissible only if the two expression sequences have the same number of expressions). Thus λ stands for a vector of eigenvalues, and U for a matrix whose columns are eigenvectors. You can check that these are indeed eigenvectors by left-multiplying each column of U in turn by **A**. For example, the following statement extracts the third column of the matrix U:

```
>  v3 := U[1..3,3]:
```

This is then multiplied by the matrix **A**:

```
>  A.v3;
```

$$\begin{bmatrix} 0.6 \\ 0.6 \\ 0.6 \end{bmatrix}$$

The vector is unchanged, confirming that `v3` is an eigenvector with eigenvalue 1.

Next, we consider the problem of finding and classifying the stationary points of the function

$$z = x^3 - 3xy + 2y^2 - 3x + 4y + 3.$$

A plot of this function is shown in Figure 2.14, which does not indicate much about the stationary points. The way to find these is to differentiate the function and use `solve`:

```
>  u := diff(z,x):  v := diff(z,y):  sol := solve({u,v});
```

$$sol := \{y = -1, x = 0\}, \{y = \frac{-7}{16}, x = \frac{3}{4}\}$$

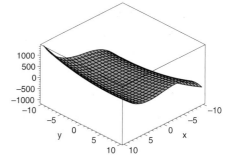

Figure 2.14   Output from `plot3d(z,x=-10..10, y=-10..10,axes=boxed)`

So the stationary points of z are $(0, -1)$ and $(3/4, -7/16)$. Note the single argument of `solve`: the two expressions u and v have only two indeterminates, namely x and y, so Maple implicitly interprets the statement as `solve({u,v},{x,y})`. All members of the `solve` family behave in this way: if the indeterminate is unambiguous, then it can be omitted. We will make use of this in the next section when solving differential equations using `dsolve`.

We could, of course, classify the stationary points of this function using the $AC - B^2$ test (you saw an example of this in *Unit 1*). However, to use a method that works in a more general context, we call on the aid of `LinearAlgebra`. We begin by calculating the Hessian matrix as follows:

> `H := <<diff(u,x),diff(u,y)>|<diff(v,x),diff(v,y)>>;`

$$H := \begin{bmatrix} 6\,x & -3 \\ -3 & 4 \end{bmatrix}$$

If you have studied MST209, you will have seen the Hessian matrix of a function of several variables defined as the matrix of second partial derivatives.

Next, consider the first stationary point `sol[1]` and let H1 be the Hessian matrix evaluated at that point:

> `H1 := eval(H,sol[1]);`

$$H1 := \begin{bmatrix} 0 & -3 \\ -3 & 4 \end{bmatrix}$$

The nature of the stationary point is given by the signs of the eigenvalues of the Hessian matrix, which we now calculate:

> `evalf(Eigenvalues(H1));`

$$\begin{bmatrix} 5.605551275 \\ -1.605551275 \end{bmatrix}$$

Table 2.2 shows how to classify the stationary point based on the signs of the eigenvalues of the Hessian matrix.

Table 2.2   Classifying stationary points (real eigenvalues)

Eigenvalues	Classification of stationary point
Both positive	Minimum
Both negative	Maximum
Opposite signs	Saddle point
Either eigenvalue zero	Test inconclusive

In this case the eigenvalues are of opposite signs; therefore the stationary point is a saddle point. The next exercise asks you to classify the other stationary point.

Exercise 2.25

Use Maple to classify the second stationary point of z.

2.5.5 Solving linear systems

Consider the problem of finding the point of intersection of three planes in three-dimensional space. Here we use a vector notation for the equations of the planes and use Maple to find the point of intersection.

The vector equation of a plane in three dimensions is given by $\mathbf{n} \cdot (\mathbf{r} - \mathbf{a}) = 0$, where $\mathbf{r}$ is the position vector of any point on the plane, $\mathbf{a}$ is a constant vector that gives one point on the plane, and $\mathbf{n}$ is a vector normal to the plane. For example, the following two vectors define a plane:

```
>    a1 := <1,2,3>:   n1 := <1,0,1>:
```

If $\mathbf{r} = x\mathbf{i} + y\mathbf{j} + z\mathbf{k}$ is a general point on the plane, then the equation of the plane is $\mathbf{n}_1 \cdot (\mathbf{r} - \mathbf{a}_1) = 0$, which in this case simplifies to $x + z = 4$. Maple can find the equation of the plane and plot it as follows:

```
>    r := <x,y,z>:   eq1 := n1.(r-a1)=0;
```
$$eq1 := x - 4 + z = 0$$

```
>    plot3d(solve(eq1,z),x=-1..1,y=-1..1,axes=boxed);
```

The `plot3d` output appears in Figure 2.15.

In addition to the plane shown in Figure 2.15, consider two further planes defined by

```
>    a2 := <2,0,9>:   n2 := <2,1,1>:
>    a3 := <3,2,5>:   n3 := <1,1,1>:
```

The equations of the three planes are three linear equations in x, y and z. If the equations are written in matrix form $\mathbf{Ar} = \mathbf{b}$, then the matrix $\mathbf{A}$ has ith row equal to $\mathbf{n}_i$, and the vector $\mathbf{b}$ has ith element $\mathbf{n}_i \cdot \mathbf{a}_i$. From the vectors $\mathbf{n}_i$ given above, it is easy to form the matrix whose columns are these $\mathbf{n}_i$, using `<n1|n2|n3>`. The matrix with these vectors as rows is obtained by transposing this matrix using `Transpose` from the `LinearAlgebra` package as follows:

```
>    A := Transpose(<n1|n2|n3>);
```
$$A := \begin{bmatrix} 1 & 0 & 1 \\ 2 & 1 & 1 \\ 1 & 1 & 1 \end{bmatrix}$$

```
>    b := <a1.n1, a2.n2, a3.n3>;
```
$$b := \begin{bmatrix} 4 \\ 13 \\ 10 \end{bmatrix}$$

To find the point of intersection, we use `LinearSolve` from the `LinearAlgebra` package, as follows:

```
>    LinearSolve(A,b);
```
$$\begin{bmatrix} 3 \\ 6 \\ 1 \end{bmatrix}$$

So the point of intersection of the three planes is $(3, 6, 1)$. Note that this point of intersection could also have been found by `solve({eq1,eq2,eq3})`, where `eq2:=n2.(r-a2)=0` and `eq3:=n3.(r-a3)=0`. The advantage of using `LinearSolve` is speed of execution, which becomes important when there are many equations and variables.

You should recall from previous study that if the scalar product of two non-zero vectors is zero, then they are orthogonal.

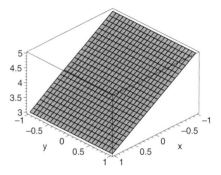

Figure 2.15 The plane $\mathbf{r} \cdot \mathbf{n}_1 = \mathbf{a}_1 \cdot \mathbf{n}_1$

There is a shorthand notation which is closer to the standard mathematical notation: `Transpose(A)` can be written `A^%T`.

Exercise 2.26

Use Maple to find the intersection of the three planes defined by $\mathbf{a}_1 = \mathbf{i} + \mathbf{j}$, $\mathbf{n}_1 = 2\mathbf{i} + \mathbf{k}$, $\mathbf{a}_2 = \mathbf{i} + \mathbf{j}$, $\mathbf{n}_2 = \mathbf{i} + \mathbf{j}$, $\mathbf{a}_3 = 0$, $\mathbf{n}_3 = \mathbf{j} + 2\mathbf{k}$.

2.6 Differential equations

This section explores some of Maple's built-in methods for solving ordinary differential equations. Subsection 2.6.1 looks at finding a general solution symbolically, i.e. in terms of known functions. Subsection 2.6.2 looks at finding particular solutions to initial-value problems symbolically. Finally, numerical solutions are studied in Subsection 2.6.3.

2.6.1 General solutions

Maple can find the symbolic solution of many different types of differential equation. For example, consider the following second-order linear homogeneous differential equation with constant coefficients:

$$\frac{d^2y}{dx^2} + 2\frac{dy}{dx} + 3y = 0. \tag{2.1}$$

How is this entered in Maple? Remembering that the differentiation command is `diff`, the first thing to try for dy/dx is `diff(y,x)`; try this now.

Exercise 2.27

Without using Maple, describe what you think the output will be when `restart: diff(y,x);` is entered into Maple. Now enter this into Maple and observe the output. Can you explain why Maple gives this output?

The above exercise shows that to represent dy/dx we need a notation to specify that y depends on x. *Unit 1* introduced the notation $y(x)$, which reads as 'y depends on x'.

With this notation, the differential equation (2.1) can be expressed as follows:

Remember that allowing y to be an *expression* that *depends* on x is *not* the same as defining y as a *function of* x.

```
>   eq := diff(y(x),x,x)+2*diff(y(x),x)+3*y(x)=0:
```

To find the general solution of the differential equation (2.1), we can use `dsolve` as follows:

```
>   sol := dsolve(eq);
```

$$sol := y(x) = \_C1\,\mathbf{e}^{(-x)}\sin(\sqrt{2}\,x) + \_C2\,\mathbf{e}^{(-x)}\cos(\sqrt{2}\,x)$$

`dsolve` is one of the `solve` family, of which `solve` itself and `fsolve` were discussed in *Unit 1*.

Notice how Maple has included the two constants that are needed to define the general solution (named $\_C1$ and $\_C2$). The automatically generated names start with an underscore so as not to conflict with any variables normally defined by the user.

After these commands have been executed, the above equation (representing the general solution of Equation (2.1)) has been assigned to the variable `sol`. There are two sides to this equation: the left-hand side, which is the expression $y(x)$, and the right-hand side, which is the more substantial (and useful) expression. For further calculations it is usually necessary to work with the latter. In this case `rhs(sol)` will yield the required expression (as you can easily check).

A more generally applicable method of obtaining the right-hand side of an equation starting $y(x) = \ldots$ is to use `eval` as follows:

```
>  eval(y(x),sol);
```

$$\_C1\, e^{(-x)} \sin(\sqrt{2}\,x) + \_C2\, e^{(-x)} \cos(\sqrt{2}\,x)$$

The `eval` command can also be used in a rather more subtle way: to check that the solution does indeed satisfy the differential equation. This is done by (in effect) substituting the solution into the differential equation, using `eval(eq,sol)` as follows:

```
>  eval(eq,sol);
```

$$0 = 0$$

The above equality is, of course, true, and signifies that the differential equation is satisfied by `sol`.

The command `dsolve` can solve many classes of differential equation, as the following exercise shows.

Exercise 2.28

Use `dsolve` to find the general solution of each of the following differential equations. Check each solution by using Maple to substitute the solution into the differential equation.

(a) $x^2 \dfrac{d^2y}{dx^2} + x\dfrac{dy}{dx} + y = 0$

(b) $\ddot{x}\cos^2 t = 1$

(c) $x\left(\dfrac{dy}{dx}\right)^2 - 2y\dfrac{dy}{dx} + 4x = 0$

Note the use of Newton's notation: for example, $\dot{x}$ is used for dx/dt, the time derivative of x.

Exercise 2.29

Use `dsolve` to find the general solution of the differential equation

$$x\dfrac{d^2y}{dx^2} + (x-1)\dfrac{dy}{dx} - y = x^2.$$

Then use `eval` to find the particular solution that satisfies $y(1) = 2$, $y'(1) = 0$.

In the next subsection you will see a more straightforward way of finding certain particular solutions such as this one.

Maple can also solve some systems of differential equations, such as

$$\dot{x} = x + y,$$
$$\dot{y} = 2x - 3y.$$

To find the general solution, give a set of differential equations to `dsolve`:

```
>  eq1 := diff(x(t),t)=x(t)+y(t):
>  eq2 := diff(y(t),t)=2*x(t)-3*y(t):
>  sol := dsolve({eq1,eq2});
```

$$sol := \{x(t) = \_C1\, e^{((-1+\sqrt{6})\,t)} + \frac{1}{2}\_C1\, e^{((-1+\sqrt{6})\,t)}\,\sqrt{6}$$
$$+ \_C2\, e^{((-1-\sqrt{6})\,t)} - \frac{1}{2}\_C2\,\sqrt{6}\,e^{((-1-\sqrt{6})\,t)},$$
$$y(t) = \_C1\, e^{((-1+\sqrt{6})\,t)} + \_C2\, e^{((-1-\sqrt{6})\,t)}\}$$

The precise form of the solution given is session-dependent. You may be given a mathematically equivalent solution with x(t) a short expression and y(t) the longer expression.

The solution is given as a set of equations. To do further computation with the expression for x(t), use `eval` to isolate the expression as follows:

```
>   xexp := eval(x(t),sol);
```

$$xexp := \_C1\,\mathbf{e}^{((-1+\sqrt{6})\,t)} + \frac{1}{2}\,\_C1\,\mathbf{e}^{((-1+\sqrt{6})\,t)}\,\sqrt{6}$$
$$+ \_C2\,\mathbf{e}^{((-1-\sqrt{6})\,t)} - \frac{1}{2}\,\_C2\,\sqrt{6}\,\mathbf{e}^{((-1-\sqrt{6})\,t)}$$

The expression `xexp` can then be used for further computations, such as collecting the exponential terms together. This can be done with the `collect` command, which takes two arguments: the expression on which `collect` operates (in this case *xexp*), and a set of expressions representing the terms to be collected (in this case $\{e^{(-1+\sqrt{6})t},\ e^{(-1-\sqrt{6})t}\}$):

```
>   collect(xexp,{exp((-1+6^(1/2))*t),exp((-1-6^(1/2))*t)});
```

$$\left(\_C1 + \frac{1}{2}\,\_C1\,\sqrt{6}\right)\,\mathbf{e}^{((-1+\sqrt{6})\,t)} + \left(\_C2 - \frac{1}{2}\,\_C2\,\sqrt{6}\right)\,\mathbf{e}^{((-1-\sqrt{6})\,t)}$$

This simplifies the form of $x(t)$.

Try the following exercise.

Use Maple to solve the system of differential equations

$$\dot{x} = 3x - 4y,$$
$$\dot{y} = 2x - y.$$

2.6.2   Initial-value problems

We assume that you know from previous study that an *initial-value problem* is a differential equation in which you are given the values of the unknown function and an appropriate number of its derivatives at a point. Maple can solve such problems, for example the following:

$$\frac{d^2y}{dx^2} + 2\frac{dy}{dx} + 3y = 0, \quad y(0) = 0,\ y'(0) = 2.$$

Initial-value problems like this can be solved using `dsolve`, once the initial values (as well as the equation) have been input; this can be done using the D operator (introduced in Section 2.1). What we do is create a set containing the differential equation together with the initial conditions. For the above example, we can input as follows:

```
>   eq := diff(y(x),x,x)+2*diff(y(x),x)+3*y(x)=0:
>   ics := y(0)=0, D(y)(0)=2:
>   sol := dsolve({eq,ics});
```

$$sol := \mathrm{y}(x) = \sqrt{2}\,e^{(-x)}\sin(\sqrt{2}\,x)$$

To plot the solution, first use `eval` to obtain the expression on the right-hand side, then use `plot` in the usual way:

```
>   yexp := eval(y(x),sol):
>   plot(yexp,x=0..10);
```

The output from `plot` is shown in Figure 2.16.

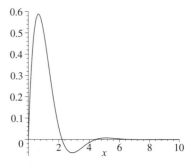

Figure 2.16 `plot` output from initial-value problem

We shall now use a simple case of rocket motion as a source of differential equations to solve, to answer such questions as 'How high will my rocket go?'. The equation of motion of a rocket, known as the **rocket equation**, can be derived using Newtonian mechanics (for completeness this derivation is included in the Appendix on page 97). The rocket equation is

$$m\dot{\mathbf{v}} - \dot{m}\mathbf{u} = \mathbf{F},$$

where the rocket has mass m and velocity $\mathbf{v}$, is acted upon by an external force $\mathbf{F}$, and emits exhaust gases with relative velocity $\mathbf{u}$.

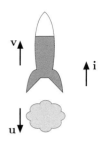

Figure 2.17   A toy rocket

Consider the toy rocket shown in Figure 2.17. When the rocket is launched, the exhaust gases exit downwards, which propels the rocket upwards. To create a simple first model, we ignore gravity and air resistance, and so assume that the external force acting on the rocket is zero, that is, $\mathbf{F} = \mathbf{0}$.

From Figure 2.17 we have $\mathbf{v} = v\,\mathbf{i}$ and $\mathbf{u} = -u\,\mathbf{i}$, where u is a positive constant and $\mathbf{i}$ is a fixed unit vector, so the $\mathbf{i}$-component of the rocket equation becomes

$$m\dot{v} + \dot{m}u = 0.$$

Now consider how to use Maple to solve this differential equation with the initial condition that the rocket starts from rest. It is important to note that the problem has two indeterminates (that is, unknowns), namely $v(t)$ and $m(t)$. We can't find the solution explicitly without more information, but we do have enough data to find $v(t)$ in terms of $m(t)$. To tell Maple to do this, we ask it to solve for $v(t)$; thus we input as follows:

```
>   eq := m(t)*diff(v(t),t)+diff(m(t),t)*u=0:
>   ic := v(0)=0:
>   sol := dsolve({eq,ic},v(t));
```
$$sol := \mathrm{v}(t) = -u\ln(\mathrm{m}(t)) + u\ln(\mathrm{m}(0))$$

The output is $v(t)$ expressed in terms of $m(t)$ and u. This equation is more conveniently written with the log terms combined, as follows:

```
>   sol := combine(sol,ln,symbolic);
```
$$sol := \mathrm{v}(t) = u\ln\left(\frac{\mathrm{m}(0)}{\mathrm{m}(t)}\right)$$

Note the use of the `symbolic` keyword to avoid having to make explicit assumptions about m.

This equation (rather than the more general one with which we started) is known as the 'rocket equation' by some authors.

In order to obtain an equation for the position $x(t)$ (that is, the height of the rocket at time t), we create a new differential equation by substituting for v:

```
>   eqx := eval(sol,v(t)=diff(x(t),t));
>   icx := x(0)=0:
```
$$eqx := \frac{d}{dt}\,\mathrm{x}(t) = u\ln\left(\frac{\mathrm{m}(0)}{\mathrm{m}(t)}\right)$$

Finally, we try to solve this differential equation:

```
>   sol := dsolve({eqx,icx},x(t));
```
$$sol := \mathrm{x}(t) = \int_0^t u\ln\left(\frac{\mathrm{m}(0)}{\mathrm{m}(\_z1)}\right)\,d\_z1$$

Since Maple outputs a definite integral, it must invent a name (in this case $\_z1$) for the variable over which the integral is expressed.

As you can see, Maple has been unable to solve this differential equation explicitly. It has however, reduced the solution to the evaluation of a definite integral. To go further and interpret this solution, a specific functional form is needed for $m(t)$. The following exercise investigates a particular $m(t)$.

Exercise 2.31

Suppose that the mass of the rocket body is M_0, the mass of the fuel is M, and the total mass of the rocket body plus fuel at time t is $m(t) = M\exp(-\gamma t) + M_0$. It is useful to know that if you input `M[0]` as part of a Maple line, the variable is output as M_0.

(a) Evaluate the solution `sol` given above with $m(t)$ as given here.

(b) The mass of the rocket body is 100 grams, the mass of the fuel is $1\,\mathrm{kg}$, the speed of the exhaust gases is $6\,\mathrm{m\,s^{-1}}$, and the rate constant γ is $10\,\mathrm{s^{-1}}$. Use Maple to plot the position of the rocket for the first two seconds of flight.

2.6.3   Numerical solutions

If a differential equation does not have a simple symbolic solution, then sometimes a numerical solution suffices. A major drawback of numerical solutions is that initial conditions must be given, so information about families of solutions of the differential equation are lost. The initial-value problem $\dot{y} + y^2 = t$, $y(0) = 1$, can be solved numerically by adding the `numeric` keyword to `dsolve`:

```
>   eq := diff(y(t),t)+y(t)^2=t:
>   ic := y(0)=1:
>   sol := dsolve({eq,ic},numeric);
```

$$sol := \mathbf{proc}(x\_rkf45) \ldots \mathbf{end\ proc}$$

The output is a procedure. You will be introduced to defining your own procedures in *Unit 3*, but here we will just *use* the procedure (procedures behave just like functions).

The output contains *rkf45*, which indicates the algorithm used to compute the solution. This is the default algorithm for the numerical solution of differential equations – the Runge–Kutta–Fehlberg fourth-fifth order algorithm, which is a general-purpose algorithm with a good balance between accuracy and speed. Other algorithms are available (see `?dsolve/numeric`).

Runge–Kutta algorithms are discussed in *MST209*.

The procedure can be used to find specific values of the solution; for example, to find the solution at $t = 1$, type

```
>   sol(1);
```

$$[t = 1., \mathrm{y}(t) = 0.833383238133406201]$$

To select the solution value from this list, note that the output is in the correct format for a list of substitutions for `eval`. So `eval(y(t),sol(1))` will evaluate to the number $0.833\,383\,238\,133\,406\,201$. This can be used to define a Maple function, say *fy*:

```
>   fy := t0->eval(y(t),sol(t0));
```

$$fy := t0 \rightarrow \mathrm{eval}(\mathrm{y}(t), \mathrm{sol}(t0))$$

Naming this function y would produce an error message as it would conflict with $y(t)$.

This function can be evaluated at numerical values to give the corresponding values of the solution curve; for example:

```
>   fy(1);
```

$$0.833383238133406201$$

This function `fy` can be used for calculations and can be plotted using `plot(fy,0..3);` or

> `plot('fy(t)',t=0..3);`

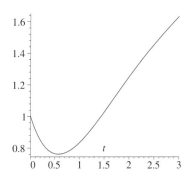

(Notice that because `fy` will evaluate only for a numerical argument, it has been surrounded by quotes as discussed in Section 2.4.) The output in Figure 2.18 shows a minimum in the interval $(0, 1)$ that we will now locate using `fsolve`. The problem needs to be stated as finding the zero of the derivative rather than minimizing a function. Rearranging the differential equation gives $\dot{y} = t - y^2$, so the location of the minimum is found by:

> `fsolve('t-fy(t)^2',t=0..1);`

$$0.5804973087$$

Figure 2.18   plot output

So the minimum occurs at $t = 0.58$ to two decimal places. Alternatively, first define an expression to be an approximation of the derivative at t using `dy:=t-fy(t)^2;`, and then use `fsolve(dy,0..1);` to find the minimum.

change to $dy := t \rightarrow t - fy(t)^2;$

We now return to an investigation of rocket motion with the next exercise.

Exercise 2.32

This exercise completes our study of rocket motion by finding an estimate of the height attained. In this exercise we add gravity to the model. In the model shown in Figure 2.17, gravity can be modelled as $\mathbf{F} = -mg\mathbf{i}$. So the equation of motion can be expressed as

$$m\ddot{x} + \dot{m}u = -mg,$$

where $g = 9.81\,\mathrm{m\,s^{-2}}$ is the magnitude of the acceleration due to gravity, and the mass of the rocket is given by $m(t) = M\exp(-\gamma t) + M_0$ using the parameters and values described in Exercise 2.31. Without loss of generality, assume that the rocket starts from rest at the origin.

The model considered previously excluded gravity and so was unable to produce an estimate. (Without gravity, the rocket just keeps going upwards.)

(a) Find the numerical solution of the equation of motion of the rocket to obtain $x(t)$, and assign the solution to the name `sol`.

(b) Use `fv:=t0->eval(diff(x(t),t),sol(t0))` to define a function `fv` that approximates the velocity of the rocket. Plot this velocity over the first two seconds of motion, and use `fsolve` to find the time taken by the rocket to reach the apex of its flight, i.e. when $v = 0$.

(c) Evaluate your solution for $x(t)$ at the time that the rocket reaches its apex to find the maximum height that the rocket attains.

2.7   *End of unit exercises*

These optional exercises show how the Maple commands introduced in this unit can be used to solve longer problems.

Exercise 2.33

The following quotation is from the novel *Mrs Miniver* by Jan Struther:

> She saw every relationship as a pair of intersecting circles. It would seem at first glance that the more they overlapped the better the relationship; but this is not so. Beyond a certain point, the law of diminishing returns sets in, and there are not enough private resources left on either side to enrich the life that is shared. Probably perfection is reached when the area of the two outer crescents, added together, is exactly equal to that of the leaf-shaped piece in the middle. On paper there must be some neat mathematical formula for arriving at this; in life, none.

In this exercise we use Maple to find the mathematical answer to Mrs Miniver's enigma. Consider the two circles in Figure 2.19. Without loss of generality, we have placed the origin at the centre of the left-hand circle, with the x-axis along the line joining the two centres. Set the length scale such that the radius of the left-hand circle is 1. Let the radius of the right-hand circle be a, and let the separation between the circles be $d > 0$.

(a) Write down the equations of the two circles. Use Maple to find the x-coordinate of the points of intersection of the circles; call this z.

(b) Write down the areas concerned as integrals, and hence use Maple to derive a relationship between a and d.

[*Hint*: You may find that `assume(a>0)` is needed in order to symbolically evaluate any integrals.]

(c) Use `implicitplot` to show the relationship between a and d.

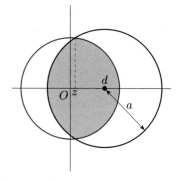

Figure 2.19   Two intersecting circles (where the shaded leaf-shaped piece should have the same area as the two outer crescents)

Exercise 2.34

You are asked to consider the problem of approximating a function f by a ratio of two polynomials p/q, which is called *rational approximation* (or Padé approximation). This is useful particularly for a function such as $\tan x$, which has discontinuities, since p/q has discontinuities at the zeros of q, and it may be possible to use these to mirror some of the discontinuities of f. One way to do this is to find an expression for the Taylor series of $fq - p$, equate it to zero, and solve for the (initially unknown) coefficients of p and q. Since $fq - p = 0$ is equivalent to $f = p/q$, the rational polynomial p/q is then the required approximation.

If p is a polynomial of order m and q is a polynomial of order n, then the approximation is called an $[m, n]$-rational approximation, and it can be shown that we need $m + n + 1$ terms of the Taylor series in order to solve for the coefficients. You are asked to find the $[3, 3]$-rational approximation of $f(x) = \tan x$ about $x = 0$, using the following steps.

(a) Let p be a general cubic polynomial and q a general cubic such that $q(0) = 1$.

(b) Use `series` to find the Taylor series of $fq - p$ of order 7 about the origin.

(c) Use `coeff` to derive equations for the coefficients of p and q, and hence use `solve` to determine the coefficients of p and q.

(d) Plot $y = \tan x$ and the resulting approximation p/q.

Requiring $q(0) = 1$ eliminates the ambiguity in the coefficients of p and q that arises because $p/q = (2p)/(2q) = \ldots$.

2.8 Appendix: The rocket equation

In this appendix, which is for interest only, we derive an equation of motion which models the behaviour of a rocket. A rocket engine converts fuel into gas molecules travelling at high speed, whose expulsion causes an alteration in the momentum of the rocket. This occurs continuously, and may be thought of as a controlled explosion. In order to model this process, we consider a two-particle system in which the mass of each particle varies while the total mass of the system remains constant.

To start with, let $t = t_0$ be a particular time at which the rocket engine is running. At this time, the system to be considered consists of the rocket and its unspent fuel. At any later time, the system is defined to consist of the rocket and unspent fuel (Particle 1) together with any gases expelled from the rocket since $t = t_0$ (Particle 2). The conversion of fuel to exhaust gas is assumed not to involve any loss of matter, so that the total mass M of the system is constant. Denoting the (variable) masses of the two particles by $m_1(t)$ and $m_2(t)$, we have

$$m_1(t) + m_2(t) = M. \tag{2.2}$$

Note that, by the definitions of the particles, we have $m_1(t_0) = M$ and $m_2(t_0) = 0$. Differentiating Equation (2.2) gives $\dot{m}_1 + \dot{m}_2 = 0$, or

$$\dot{m}_2 = -\dot{m}_1, \tag{2.3}$$

expressing the fact that the rate of mass loss for Particle 1 (the 'rocket' particle) is equal to the rate of mass gain for Particle 2 (the 'gases' particle). The system is illustrated in Figure 2.20, where the velocities of the two particles at time t are denoted by $\mathbf{v}_1(t)$ and $\mathbf{v}_2(t)$, respectively.

Apply Newton's second law in the form $\dot{\mathbf{P}} = \mathbf{F}$, where $\mathbf{P} = m_1\mathbf{v}_1 + m_2\mathbf{v}_2$ is the total momentum of the system, and $\mathbf{F}$ is the total external force. This produces the equation

$$\dot{m}_1\mathbf{v}_1 + m_1\dot{\mathbf{v}}_1 + \dot{m}_2\mathbf{v}_2 + m_2\dot{\mathbf{v}}_2 = \mathbf{F},$$

which from Equation (2.3) can be rewritten as

$$m_1\dot{\mathbf{v}}_1 - \dot{m}_1(\mathbf{v}_2 - \mathbf{v}_1) + m_2\dot{\mathbf{v}}_2 = \mathbf{F}. \tag{2.4}$$

Thus far we have a differential equation which holds at or after time $t = t_0$. The next step is to focus on the time $t = t_0$ itself. From the definition of the particles we have $m_2(t_0) = 0$. Writing $\mathbf{u}$ for the velocity of the exhaust gases relative to the rocket at the moment of expulsion, we also have $\mathbf{u}(t_0) = \mathbf{v}_2(t_0) - \mathbf{v}_1(t_0)$. Thus at $t = t_0$, Equation (2.4) gives

$$m_1(t_0)\dot{\mathbf{v}}_1(t_0) - \dot{m}_1(t_0)\mathbf{u}(t_0) = \mathbf{F}(t_0).$$

Since the 'gases' particle has not yet materialized at time $t = t_0$, the right-hand side of this equation must be the external force acting on the 'rocket' particle alone. The time $t = t_0$ was chosen arbitrarily, so the motion of the 'rocket' particle should satisfy the equation at any time while its engine is running. Finally, since we no longer have any explicit reference to Particle 2, we can simplify matters by removing the subscript 1 throughout. This leaves the equation of motion known as the **rocket equation**,

$$m\dot{\mathbf{v}} - \dot{m}\mathbf{u} = \mathbf{F},$$

where the rocket has mass m and velocity $\mathbf{v}$, is acted upon by an external force $\mathbf{F}$, and emits exhaust gases with relative velocity $\mathbf{u}$.

You may find it easier to visualize the system as being made up of two subsystems of particles rather than two particles. The mathematical argument is the same in either case, since each subsystem may be represented by a particle at its centre of mass.

(a) At time t_0:

$\mathbf{v}_1(t_0)$

$m_1(t_0) = M$

(b) At time t:

$\mathbf{v}_2(t)$ $\mathbf{v}_1(t)$

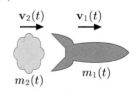

$m_2(t)$ $m_1(t)$

Figure 2.20 (a) The situation at time $t = t_0$ (b) The state of the system at some later time t

Writing $\dot{\mathbf{v}}$ as the acceleration, $\mathbf{a}$, of the rocket, this equation can be seen as extending $\mathbf{F} = m\mathbf{a}$ to apply to variable-mass particles.

Summary of Unit 2

The main commands introduced in this unit are as follows.

`->`	for defining functions
`[n]`	for indexing expression sequences, lists, strings, vectors, etc.
`seq`	for constructing an expression sequence
`select`	for selecting elements from a list or set
`map`	for transforming elements from a list or set
`series`	for constructing a series from an expression
`Eigenvectors`	for solving eigenvalue problems
`LinearSolve`	for solving systems of linear equations
`dsolve`	for solving differential equations

Learning outcomes

After studying this unit you should be able to:

- understand and use the following Maple data types: expression sequence, list, set, `Vector` and `Matrix`;
- understand and use indexing of expression sequences, lists, sets, vectors and matrices;
- understand and use the following Maple facilities: `seq`, `nops`, `op`, `select`, `map`, `||`, `convert`, `->`, `@`, `@@`, `D`;
- use `Eigenvectors` to solve eigenvalue problems;
- use `LinearSolve` to solve linear systems of algebraic equations;
- use `dsolve` to find symbolic solutions of differential equations, systems of differential equations and initial-value problems;
- use `dsolve` to find numerical solutions of initial-value problems;
- produce graphs using `plot`, `plot3d`, `Statistics[ColumnGraph]`, `contourplot` and `implicitplot`.

Solutions to Exercises

Solution 2.1

The Maple function is defined as follows:

```
>  f := x->(x-1)^2/2;
```

$$f := x \to \frac{1}{2}(x-1)^2$$

The function is evaluated as follows:

```
>  f(-1),f(0),f(1);
```

$$2, \frac{1}{2}, 0$$

Solution 2.2

(a) The function $f(x, y) = x^y$ is defined using

```
>  f := (x,y)->x^y;
```

$$f := (x, y) \to x^y$$

As a check, try to compute 2^3 using the function:

```
>  f(2,3);
```

$$8$$

(b) The function $z = \sin x \cos y$ is defined using

```
>  z := (x,y)->sin(x)*cos(y);
```

$$z := (x, y) \to \sin(x)\cos(y)$$

As a check, evaluate at the point $(\pi/3, \pi/3)$:

```
>  z(Pi/3,Pi/3);
```

$$\frac{\sqrt{3}}{4}$$

The value is $\frac{\sqrt{3}}{2} \times \frac{1}{2}$, as expected.

(c) The function $w = u^3 + 2uv + v^2$ is defined using

```
>  w := (u,v)->u^3+2*u*v+v^2;
```

$$w := (u, v) \to u^3 + 2uv + v^2$$

The function value at the point $(1, 2)$ is $1^3 + 2 \times 1 \times 2 + 2^2 = 1 + 4 + 4 = 9$, and this can be used to check the Maple output:

```
>  w(1,2);
```

$$9$$

Solution 2.3

(a) Define f to be the function $x \longmapsto x^2 + 1$; then the function $\sin(x^2 + 1)$ can be written as a composite g:

```
>  f := x->x^2+1;  g := sin@f;
```

$$f := x \to x^2 + 1$$
$$g := \sin@f$$

Check:

```
>  g(Pi),g(sqrt(Pi));
```

$$\sin(\pi^2 + 1), -\sin(1)$$

(b) Define f to be the function $y \longmapsto y^2 + 1$ and g to be the composite:

```
>  f := y->y^2+1;  g := f@cos;
```

$$f := y \to y^2 + 1$$
$$g := f@\cos$$

Check:

```
>  g(0),g(Pi/2);
```

$$2, 1$$

(c) Define a function f to be the square of its argument and g to be the composite:

```
>  f := x->x^2;  g := f@ln;
```

$$f := x \to x^2$$
$$g := f@\ln$$

Check:

```
>  g(3);
```

$$\ln(3)^2$$

(d) Define f as in the previous part and g as the composite with the functions applied in the reverse order:

```
>  f := x->x^2;  g := ln@f;
```

$$f := x \to x^2$$
$$g := \ln@f$$

Check:

```
>  g(3);
```

$$2\ln(3)$$

Note that the result is different from the result in part (c) – be careful with the order in which functions are composed.

Solution 2.4

The function $f(x) = x/2 + 1/x$ is defined in Maple as follows:

```
>  f := x->x/2+1/x;
```

$$f := x \to \frac{1}{2}x + \frac{1}{x}$$

Now use the repeated composition operator @@ to apply this function repeatedly. Start with a floating-point number in order to give a floating-point result (otherwise you will see some very big fractions).

```
>  (f@@20)(1.0);
```

$$1.414213562$$

Try starting with a different number:

```
>  (f@@20)(2.0);
```

$$1.414213562$$

The result always seems to be the same constant for any positive starting number. (The constant is an approximation to $\sqrt{2}$. This result is a consequence of the Newton–Raphson method for solving equations – a method that you will meet later in the course.) Starting with zero gives a division by zero error, and starting with a negative number converges to $-\sqrt{2}$.

Solution 2.5

(a) Calculate the derived function as follows:

```
>  y := x->sin(cos(x)):  D(y);
```
$$x \to -\cos(\cos(x))\sin(x)$$

Note that the output is a function and not an expression.

(b) The derived function can be calculated by:

```
>  z := t->(1+t)^3:  D(z);
```
$$t \to 3\left(1+t\right)^2$$

Solution 2.6

First find the solution:

```
>  sol := solve(sinh(2*x)-3=y,x):
```

Then use **unapply** to define the function h:

```
>  h := unapply(sol,y);
```
$$h := y \to \frac{1}{2}\operatorname{arcsinh}(3+y)$$

Now compute $h(0)$:

```
>  h(0.0);
```
$$0.9092232295$$

Solution 2.7

(a) The ith even number is $2i$, so the sequence of the first five even numbers is created by:

```
>  seq(2*i,i=1..5);
```
$$2,\, 4,\, 6,\, 8,\, 10$$

(b) The sequence consisting of the first ten odd numbers can be constructed using a formula for the ith odd number as follows:

```
>  seq(2*i-1,i=1..10);
```
$$1,\, 3,\, 5,\, 7,\, 9,\, 11,\, 13,\, 15,\, 17,\, 19$$

(c) Using the hint, the sequence of the first ten prime numbers can be generated by:

```
>  seq(ithprime(i),i=1..10);
```
$$2,\, 3,\, 5,\, 7,\, 11,\, 13,\, 17,\, 19,\, 23,\, 29$$

(d) The sequence of powers of x can be generated by:

```
>  seq(x^i,i=0..5);
```
$$1,\, x,\, x^2,\, x^3,\, x^4,\, x^5$$

(e) Using the given formula generates the sequence:

```
>  seq(sin(n*Pi/2),n=0..9);
```
$$0,\, 1,\, 0,\, -1,\, 0,\, 1,\, 0,\, -1,\, 0,\, 1$$

Solution 2.8

(a) The given function can be defined by:

```
>  fs := (a,b)->(b,a+b);
```
$$fs := (a,\, b) \to (b,\, a+b)$$

(b) The function **fs** performs one step along the Fibonacci sequence. If a and b are the latest two terms of the sequence, then the result is the second of these, b, and the next term, a+b. For example, the sequence starts $1, 1$, so we compute:

```
>  fs(1,1);
```
$$1,\, 2$$

The result is the next pair of terms in the sequence, 1 and 2. So to calculate the 100th term we need to apply **fs** to $(1,1)$ a total of 98 times (using the repeated composition operator @@):

```
>  (fs@@98)(1,1);
```
$$218922995834555169026,\, 354224848179261915075$$

So the 100th term in the Fibonacci sequence is $354\,224\,848\,179\,261\,915\,075$.

Solution 2.9

The given list is defined by the following:

```
>  L := [seq(i,i=1..5)];
```
$$L := [1,\, 2,\, 3,\, 4,\, 5]$$

When defining functions it is sometimes easier to experiment with the body of the function in order to get this correct. In this case the body of the function uses **seq** with a step of -1:

```
>  [seq(L[i],i=nops(L)..1,-1)];
```
$$[5,\, 4,\, 3,\, 2,\, 1]$$

So the function **reverse** can be defined by

```
>  reverse := L->[seq(L[i],i=nops(L)..1,-1)]:
```

Applying this function reverses the list:

```
>  reverse(L);
```
$$[5,\, 4,\, 3,\, 2,\, 1]$$

Solution 2.10

A first attempt at creating the amalgamated list is to try **join:=(x,y)->[x,y]**, but this adds extra pairs of square brackets to the list, i.e. the result is a list consisting of two lists. What is required is to use **op** to remove the extra square brackets, as follows:

```
>  join := (x,y)->[op(x),op(y)]:
```

Solution 2.11

(a) The strategy is to first construct a list of numbers in the required range, and then select only the prime numbers:

```
>  L := [seq(i,i=101..199)]:
>  select(isprime,L);
```
$$[101,\, 103,\, 107,\, 109,\, 113,\, 127,\, 131,\, 137,\, 139,\, 149,\, 151,$$
$$157,\, 163,\, 167,\, 173,\, 179,\, 181,\, 191,\, 193,\, 197,\, 199]$$

(b) First construct the list of all solutions of the equation:

```
>  sol := [solve(x^4+x^2+1,x)]:
```

Now use **select** to obtain only the solutions with positive real part, i.e. in the right-hand half of the complex plane:

```
>  realPartPos := x->Re(x)>0:
>  select(realPartPos,sol);
```
$$\left[\frac{1}{2}+\frac{1}{2}I\sqrt{3},\, \frac{1}{2}-\frac{1}{2}I\sqrt{3}\right]$$

(c) First construct a list of all solutions:

```
> Lsol :=
  [solve({x+3*y=1,x^2+y^2=1},{x,y})];
```

$$Lsol := \left[\{x = 1, y = 0\}, \{y = \frac{3}{5}, x = \frac{-4}{5}\}\right]$$

Now use `select` to remove the solution with $y = 0$:

```
> nonzero := sol->eval(y,sol)<>0:
> select(nonzero,Lsol);
```

$$\left[\{y = \frac{3}{5}, x = \frac{-4}{5}\}\right]$$

Solution 2.12

(a) Start with the list of prime numbers:

```
> L100 := [seq(ithprime(n),n=1..100)]:
```

(b) Now use `select` to find only the twin primes:

```
> twin := p->isprime(p+2):
> select(twin,L100);
```

$$[3, 5, 11, 17, 29, 41, 59, 71, 101, 107, 137, 149,$$
$$179, 191, 197, 227, 239, 269, 281, 311, 347,$$
$$419, 431, 461, 521]$$

Solution 2.13

Start with the given list:

```
> L := [seq(i,i=0..4)];
```

$$L := [0, 1, 2, 3, 4]$$

(a) The list of odd numbers can be obtained by using `map` with the following function:

```
> odds := n->2*n+1:
> map(odds,L);
```

$$[1, 3, 5, 7, 9]$$

(b) The list of powers of x can be obtained by using `map` with the following function:

```
> powx := n->x^n:
> map(powx,L);
```

$$[1, x, x^2, x^3, x^4]$$

Solution 2.14

(a) The set of perfect powers can be obtained by using two `seq` statements to create an expression sequence. Converting the expression sequence to a set automatically removes any repeated elements:

```
> M := 1000:  N := 10:
> P := {seq(seq(m^n,m=2..M),n=2..N)}:
```

(b) Use `add` to compute the first sum as follows:

```
> add(1.0/(P[i]-1),i=1..nops(P));
               0.9983859147
```

Note that the order of the elements of a set is not fixed, so the notation `P[i]` needs to be used with care if P is a set. In this case the use is permissible because the value of the overall command is not affected by the order of the elements of P.

(The above value is close to 1 and you might like to check that it becomes closer to 1 as more terms are added to the summation. The theorem that the sum of this series is 1 is known as the Goldbach–Euler theorem.)

The double sum mentioned in the question can be computed by:

```
> add(add(1.0/(m^n-1),m=2..M),n=2..N);
               1.128410752
```

This gives a different value from the sum over the set of perfect powers. The reason for the discrepancy is that the duplicate elements of the sequence have been eliminated in P. For example, 16 would be counted both as 2^4 and as 4^2, and so would contribute more to the second sum than to the first.

Solution 2.15

(a) To find the series up to and including the x^7 term, we need to truncate at 8 in the `series` command:

```
> S := series(ln(1+x-sin(x)),x=0,8);
```

$$S := \frac{1}{6} x^3 - \frac{1}{120} x^5 - \frac{1}{72} x^6 + \frac{1}{5040} x^7 + O(x^8)$$

(b) These expansions need to be converted to polynomials and then one subtracted from the other, as follows:

```
> sc := convert(series(2*sinh(x)*cosh(x),
        x=0,8),polynom);
```

$$sc := 2x + \frac{4}{3} x^3 + \frac{4}{15} x^5 + \frac{8}{315} x^7$$

```
> s2 := convert(series(sinh(2*x),
        x=0,8),polynom);
```

$$s2 := 2x + \frac{4}{3} x^3 + \frac{4}{15} x^5 + \frac{8}{315} x^7$$

```
> d := sc-s2;
```

$$d := 0$$

Alternatively, find the series of the difference:

```
> series(2*sinh(x)*cosh(x)-sinh(2*x),x=0,8);
```

$$O(x^8)$$

The presence of nothing but the order term indicates that the series is zero up to and including terms in x^7.

(c) In this case the expansion is:

```
> S := series(ln(1+sqrt(x)+x),x=0,4);
```

$$S := \sqrt{x} + \frac{x}{2} - \frac{2 x^{(3/2)}}{3} + \frac{x^2}{4}$$
$$+ \frac{x^{(5/2)}}{5} - \frac{x^3}{3} + \frac{x^{(7/2)}}{7} + O(x^4)$$

Note that this is not a Taylor series, because it contains odd powers of $\sqrt{x}$.

The exercise asks for the expansion 'up to and including the term in x^3', so (strictly speaking) the term $\dfrac{x^{(7/2)}}{7}$ should be discarded.

(d) Finally, the expansion for $\cot x$ is found as follows:

```
> S := series(cot(x),x=0,4);
```

$$S := x^{-1} - \frac{1}{3} x - \frac{1}{45} x^3 + O(x^4)$$

Again, this is not a Taylor series, this time because of the x^{-1} term.

Solution 2.16

The function can be plotted as follows:

```
>   f := x->mul(k^(-1/4),k=1..x):
>   plot(f,1..10);
```

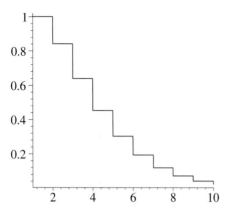

Solution 2.17

The vertices of the given regular pentagon are equally spaced around the unit circle, so the angle between one vertex and the next is $2\pi/5$ radians. The coordinates of the vertices can be computed by:

```
>   pent :=
    [seq([cos(2*Pi*i/5),sin(2*Pi*i/5)],
    i=0..5)]:
```

(The range is 0..5 rather than 0..4 in order that the last point of the pentagon is joined to the first point.)

The pentagon is drawn using plot:

```
>   plot(pent,-1..1,-1..1);
```

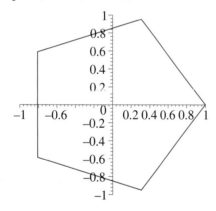

Note the explicit ranges for the x- and y-axes so that the regular pentagon will not be distorted.

Solution 2.18

A 3D plot of the function is obtained by:

```
>   f := x*(x^2-3*y^2)*exp(1-x^2-y^2):
>   plot3d(f,x=-4..4,y=-4..4,axes=boxed);
```

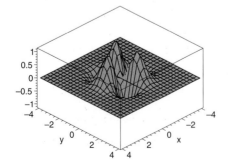

A contour plot is produced by:

```
>   plots[contourplot](f,x=-4..4,y=-4..4);
```

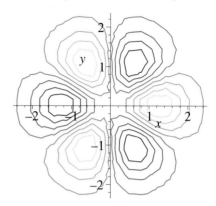

From the contour plot there are seven stationary points: three maxima and three minima, and a saddle point at the origin. One stationary point is in the region with $1 \le x \le 2$ and $-0.5 \le y \le 0.5$, and we can use fsolve to locate it:

```
>   sol := fsolve({diff(f,x),diff(f,y)},
                  {x=1..2,y=-0.5..0.5});
```

$$sol := \{x = 1.224744871, y = 0.\}$$

Now use fsolve to find the stationary point located in the region with $0 \le x \le 1$ and $0.5 \le y \le 1.5$:

```
>   sol := fsolve({diff(f,x),diff(f,y)},
                  {x=0..1,y=0.5..1.5});
```

$$sol := \{x = 0.6123724357, y = 1.060660172\}$$

Four other stationary points can be found similarly, or you could use symmetry, to find the locations as $(\pm1.22, 0)$, $(\pm0.61, \pm1.06)$ to two decimal places.

To demonstrate that the stationary points not at the origin lie on a circle centred at the origin, we calculate the distance of the second stationary point from the origin:

```
>   eval(sqrt(x^2+y^2),sol);
```

$$1.224744871$$

This is the same as the distance of the points $(\pm1.22, 0)$ from the origin (and also the other points $(\pm0.61, \pm1.06)$). So all six stationary points not at the origin lie on the circle with the origin as centre and radius $1.224\,744\,871$.

Solution 2.19

First plot the relationship using implicitplot:

```
>   eq := tanh(x+y)*(1+3*sin(y)^2)=x:
>   plots[implicitplot](eq,x=-4..4,y=-4..4);
```

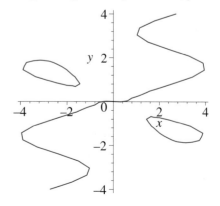

From the picture, the largest range for which the mapping from x to y is single-valued that includes the origin is $(-1, 1)$. This can be found by clicking on the picture to obtain approximate values for the closest approach of the several branches of the mapping to the y-axis.

(Alternatively, use `fsolve` to locate the turning points of the function as follows:

```
> fsolve({eq,diff(eq,y)},{x=0..2,y=2..4});
```
$$\{x = 0.9994940692, y = 3.141423939\}$$

This gives the closest approach of the branch of the curve through the origin at its first turning point for x as a function of y with x positive. Compare this with the closest approach of the closed branch of the curve:

```
> fsolve({eq,diff(eq,y)},{x=0..2,y=-1..0});
```
$$\{x = 1.444079519, y = -0.7964293839\}$$

This x-value is greater than 1, so the former turning point places the greater restriction on the range in which the function is single-valued.)

Solution 2.20

(a) The following commands will allow you to view a regular polygon of any order:

```
> with(plots):
```
```
> poly := N->
    [seq([cos(2*Pi*i/N),sin(2*Pi*i/N)],i=0..N)]:
```
```
> plotpoly := N->plot(poly(N),-1..1,-1..1,
    title="Regular "||N||"-gon"):
```

You can now view a regular polygon of order (say) 6 by typing `plotpoly(6);`. (To save space, the plot is not shown here.)

(b) To create an animation, add the following:

```
> polyseq := seq(plotpoly(N),N=3..22):
```
```
> display(polyseq,insequence=true);
```

Solution 2.21

The most efficient way to do this is to define the polynomial as a function, then set up and solve three simultaneous equations:

```
> p := x->add(a||i*x^i,i=0..2):
```
```
> eq := p(1)=2.5, p(2)=1.5, p(3)=Pi:
```
```
> sol := solve({eq}):
```
```
> px := eval(p(x),sol); plot(px,x=0..4);
```
$$px := 6.141592654 - 4.962388980\,x + 1.320796327\,x^2$$

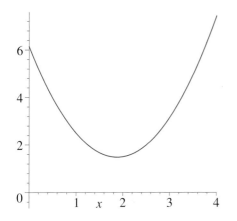

Solution 2.22

(a) The vector $3\mathbf{i} + 2\mathbf{j} + 5\mathbf{k}$ is given by

```
> <3,2,5>;
```
$$\begin{bmatrix} 3 \\ 2 \\ 5 \end{bmatrix}$$

Alternatively, the vector can be created from a list:

```
> Vector([3,2,5]):
```

(b) A vector of the first five even numbers is given by

```
> Vector(5,i->2*i);
```
$$\begin{bmatrix} 2 \\ 4 \\ 6 \\ 8 \\ 10 \end{bmatrix}$$

Solution 2.23

(a) The given matrix can be created by

```
> Matrix([[2,0,9],[3,2,5]]);
```
$$\begin{bmatrix} 2 & 0 & 9 \\ 3 & 2 & 5 \end{bmatrix}$$

An alternative is `<<2|0|9>,<3|2|5>>`.

(b) To create the given matrix without explicitly working out all of the elements by hand, create a list using `seq` and the given expression, and then convert the list into a matrix:

```
> L := [seq([seq(abs(i-j),i=1..3)],j=1..3)]:
```
```
> Matrix(L);
```
$$\begin{bmatrix} 0 & 1 & 2 \\ 1 & 0 & 1 \\ 2 & 1 & 0 \end{bmatrix}$$

Alternatively, the given matrix can be created by an initializer function (as described for creating vectors):

```
> Matrix(3,3,(i,j)->abs(i-j)):
```

Solution 2.24

The following statements use the given function to calculate the product of $\mathbf{v}$ by $\mathbf{A}$ one hundred times:

```
> v := <1,0,0>:
```
```
> A := <<0.3,0.2,0.5>|<0.2,0.6,0.2>|
    <0.5,0.2,0.3>>:
```
```
> mulA := v->A.v:
```
```
> (mulA@@100)(v);
```
$$\begin{bmatrix} 0.333333333333333370 \\ 0.333333333333333370 \\ 0.333333333333333370 \end{bmatrix}$$

If we let $\mathbf{u}$ represent this output, then multiply again by $\mathbf{A}$, the result is unchanged; that is, $\mathbf{Au} = \mathbf{u}$:

```
> u := %:  A.u;
```
$$\begin{bmatrix} 0.333333333333333370 \\ 0.333333333333333370 \\ 0.333333333333333370 \end{bmatrix}$$

This means that $\mathbf{u}$ is an eigenvector of $\mathbf{A}$ with eigenvalue 1.

Solution 2.25

The procedure is exactly the same as in the text. First use Maple to find the Hessian matrix at the second stationary point:

```
>  H2 := eval(H,sol[2]);
```

$$H2 := \begin{bmatrix} \dfrac{9}{2} & -3 \\ -3 & 4 \end{bmatrix}$$

Now compute the eigenvalues:

```
>  evalf(Eigenvalues(H2));
```

$$\begin{bmatrix} 7.260398645 \\ 1.239601355 \end{bmatrix}$$

Both of the eigenvalues are positive, so the stationary point is a local minimum.

Solution 2.26

Proceed as in the text by first defining the vectors in Maple:

```
>  a1 := <1,1,0>:   n1 := <2,0,1>:
>  a2 := <1,1,0>:   n2 := <1,1,0>:
>  a3 := <0,0,0>:   n3 := <0,1,2>:
```

Now form the matrix **A** and the vector **b**:

```
>  A := Transpose(<n1|n2|n3>);
   b := <a1.n1,a2.n2,a3.n3>;
```

$$A := \begin{bmatrix} 2 & 0 & 1 \\ 1 & 1 & 0 \\ 0 & 1 & 2 \end{bmatrix}$$

$$b := \begin{bmatrix} 2 \\ 2 \\ 0 \end{bmatrix}$$

Now use `LinearSolve` to find the point of intersection:

```
>  LinearAlgebra[LinearSolve](A,b);
```

$$\begin{bmatrix} \dfrac{6}{5} \\ \dfrac{4}{5} \\ \dfrac{-2}{5} \end{bmatrix}$$

So the point of intersection of the three planes is $(6/5, 4/5, -2/5)$.

Solution 2.27

The Maple output is 0. To understand why this is the output, recall that `diff` gives the partial derivative and that all expressions that do not explicitly depend on x are assumed constant. In this case y is assumed to be a constant, so the derivative is zero.

Solution 2.28

(a) The solution can be obtained as follows:

```
>  eq := x^2*diff(y(x),x,x)+x*diff(y(x),x)
         +y(x)=0:
>  sol := dsolve(eq);
```

$$y(x) = \_C1 \sin(\ln(x)) + \_C2 \cos(\ln(x))$$

As a check, we evaluate the equation eq using the solution sol and simplify:

```
>  chk := eval(eq,sol):  simplify(chk);
```

$$0 = 0$$

Thus this solution satisfies the equation.

(b) This solution can be generated as follows:

```
>  eq := diff(x(t),t,t)*cos(t)^2=1:
>  sol := dsolve(eq);
```

$$x(t) = -\ln(\cos(t)) + \_C1\,t + \_C2$$

Check:

```
>  chk := eval(eq,sol):  simplify(chk);
```

$$1 = 1$$

So this solution satisfies the equation.

(c) The solution is obtained as follows:

```
>  eq := x*diff(y(x),x)^2-2*y(x)*diff(y(x),x)
         +4*x=0:
>  sol := dsolve(eq);
```

$$sol := y(x) = 2\,x,\ y(x) = -2\,x,\ y(x) = \left(\frac{x^2}{2\,\_C1^2} + 2\right)\_C1$$

```
>  chk := eval(eq,sol[3]):  simplify(chk);
```

$$0 = 0$$

In this case Maple has returned three solutions: two exceptional solutions, followed by a family of solutions. The first two solutions obviously satisfy the equation, so here we check the third solution:

```
>  chk := eval(eq,sol[3]):  simplify(chk);
```

$$0 = 0$$

Solution 2.29

The solution is (as usual) found using `dsolve`:

```
>  eq := x*diff(y(x),x,x)
         +(x-1)*diff(y(x),x)-y(x)=x^2:
>  sol := dsolve(eq);
```

$$sol := y(x) = (x - 1)\_C2 + \mathbf{e}^{(-x)}\_C1 + x^2$$

The first step in finding the particular solution is to define two Maple equations which correspond to the initial conditions, as follows:

```
>  ye := eval(y(x),sol):
>  eq1 := eval(ye=2,x=1);
   eq2 := eval(diff(ye,x)=0,x=1);
```

$$eq1 := 1 + \mathbf{e}^{(-1)}\_C1 = 2$$

$$eq2 := \_C2 - \mathbf{e}^{(-1)}\_C1 + 2 = 0$$

To find the particular solution, use `solve` to find the constants $\_C1$ and $\_C2$ and substitute these into the solution:

```
>  Csol := solve({eq1,eq2}):
>  eval(sol,Csol);
```

$$y(x) = -x + 1 + \mathbf{e}^{(-x)}\,\mathbf{e} + x^2$$

So the particular solution is

$$y = \exp(1 - x) + x^2 - x + 1.$$

Solution 2.30

The system is solved as follows:

```
>  eq1 := diff(x(t),t)=3*x(t)-4*y(t):
>  eq2 := diff(y(t),t)=2*x(t)-y(t):
>  sol := dsolve({eq1,eq2});
```

$$sol := \{\mathrm{x}(t) = \mathbf{e}^t \left(\_C1 \sin(2\,t) + \_C2 \cos(2\,t) \right),$$
$$y(t) = \frac{1}{2}\, \mathbf{e}^t \left(\_C1 \sin(2\,t) - \_C1 \cos(2\,t) \right.$$
$$\left. + \_C2 \cos(2\,t) + \_C2 \sin(2\,t) \right)\}$$

Solution 2.31

(a) Start by repeating the steps given in the text to find the solution `sol`:

```
>   eq := m(t)*diff(v(t),t)+diff(m(t),t)*u=0:
>   ic := v(0)=0:
>   sol := dsolve({eq,ic},v(t)):
>   sol := combine(sol,ln,symbolic):
>   eqx := eval(sol,v(t)=diff(x(t),t)):
>   icx := x(0)=0:
>   sol := dsolve({eqx,icx},x(t));
```

$$sol := \mathrm{x}(t) = \int_0^t u \ln\left(\frac{\mathrm{m}(0)}{\mathrm{m}(\_z1)} \right) d\_z1$$

Now define m as a function and re-evaluate `sol`:

```
>   m := t->M*exp(-gamma*t)+M[0]:
>   sol;
```

$$sol := \mathrm{x}(t) = \int_0^t u \ln\left(\frac{M + M_0}{M\, \mathbf{e}^{(-\gamma\,\_z1)} + M_0} \right) d\_z1$$

This is the required solution.

(b) The given parameters can be entered as a list:

```
>   param := [M[0]=0.1, M=1, gamma=10, u=6]:
```

This list can be used to substitute into `sol`. Note the ordering of the list – M_0 must appear before M in the list otherwise an error will result (Maple would arrive at 1_0 after the M substitution).

```
>   xexp := eval(rhs(sol),param);
```

$$xexp := \int_0^t 6 \ln\left(\frac{1.1}{\mathbf{e}^{(-10\,\_z1)} + 0.1} \right) d\_z1$$

This expression `xexp` now contains no free parameters (apart from t) and so it can be plotted in the usual way:

```
>   plot(xexp,t=0..2);
```

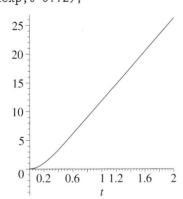

Observe that the rocket propulsion takes effect only over the first 0.2 seconds of the motion. After this time the curve is a straight line, which is expected since no external forces are acting on the rocket.

Solution 2.32

(a) Write the given equation in Maple:

```
>   m := t->M*exp(-gamma*t)+M[0]:
>   eq := m(t)*diff(x(t),t,t)
            +diff(m(t),t)*u=-m(t)*g;
```

$$eq := (M\, \mathbf{e}^{(-\gamma\,t)} + M_0) \left(\frac{d^2}{dt^2}\, \mathrm{x}(t) \right) - M\,\gamma\, \mathbf{e}^{(-\gamma\,t)}\, u$$
$$= -(M\, \mathbf{e}^{(-\gamma\,t)} + M_0)\, g$$

Use `eval` to substitute values for the free parameters:

```
>   params :=
      [M[0]=0.1,M=1,gamma=10,u=6,g=9.81]:
>   eq := eval(eq,params);
```

$$eq := (\mathbf{e}^{(-10\,t)} + 0.1) \left(\frac{d^2}{dt^2}\, \mathrm{x}(t) \right) - 60\, \mathbf{e}^{(-10\,t)}$$
$$= -9.81\, \mathbf{e}^{(-10\,t)} - 0.981$$

Now find the numerical solution of the differential equation using the given initial conditions:

```
>   ics := D(x)(0)=0,x(0)=0:
>   sol := dsolve({eq,ics},numeric);
```

$$sol := \mathbf{proc}(x\_rkf45) \ldots \mathbf{end\ proc}$$

(b) Using the function definition given in the exercise, the velocity of the rocket can be plotted as follows:

```
>   fv := t0->eval(diff(x(t),t),sol(t0)):
>   plot(fv,0..2);
```

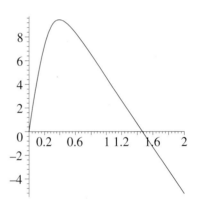

From the graph we see that the first time the rocket is stationary after the flight begins is located somewhere between $t = 1$ and $t = 2$. Use `fsolve` to find the exact moment, and assign this to `ta`:

```
>   ta := fsolve(fv,1..2);
```
$$ta := 1.466600284$$

(c) All that remains is to evaluate the numerical solution at the time `ta`:

```
>   fx := t0->eval(x(t),sol(t0)):
>   fx(ta);
```
$$8.03131775445005225$$

So the toy rocket is expected to reach a height of about 8 metres.

Solution 2.33

(a) The two circles have equations $x^2 + y^2 = 1$ and $(x - d)^2 + y^2 = a^2$, and we enter these into Maple:

```
>   eq1 := x^2+y^2=1:
>   eq2 := (x-d)^2+y^2=a^2:
```

Now use `solve` to find the points of intersection:

```
>   sol := solve({eq1,eq2},{x,y}):
```

Define z to be the x-coordinate of the points of intersection:

```
>   z := eval(x,sol);
```

$$z := \frac{d^2 + 1 - a^2}{2\,d}$$

(This result is easy to derive by hand by subtracting eq1 from eq2.)

(b) Let A be the area of the shaded region in Figure 2.19. Calculate A as twice the area from the x-axis to the positive parts of the arcs of circles (using the hint to help Maple evaluate the integrals):

```
>   assume(a>0):

>   A := 2*int(sqrt(a^2-(x-d)^2),x=d-a..z)
        +2*int(sqrt(1-x^2),x=z..1):
```

The area of the left-hand crescent is the area of the whole circle minus the shaded area, i.e. $\pi - A$. Similarly, the area of the right-hand crescent is $\pi a^2 - A$. So the condition that the shaded area is equal to the sum of the two crescents is $A = (\pi - A) + (\pi a^2 - A)$. This simplifies to $3A - \pi - \pi a^2 = 0$, and this is the condition entered into Maple:

```
>   eq := simplify(3*A-Pi-Pi*a^2=0,symbolic);
```

$$eq := \frac{a{\sim}^2 \pi}{2} - 3\,a{\sim}^2 \arcsin\left(\frac{d^2 - 1 + a{\sim}^2}{2\,d\,a{\sim}}\right)$$
$$- \frac{3\sqrt{2\,d^2\,a{\sim}^2 - d^4 + 2\,d^2 - 1 + 2\,a{\sim}^2 - a{\sim}^4}}{2}$$
$$- 3\arcsin\left(\frac{d^2 + 1 - a{\sim}^2}{2\,d}\right) + \frac{\pi}{2} = 0$$

Recall that the trailing tilde ($\sim$) after a in the above expression is a reminder that the assumption is being made that a is positive.

(c) We wish to plot the relationship between a and d with `implicitplot`. We could just experiment with various ranges, but it is useful to start by asking what ranges we shall require.

The range for d is reasonably clear; for the right-hand circle to have its centre to the right of the centre of the other circle but within that circle, we must specify $0 < d < 1$. Now, if either circle has more than twice the area of the other, then (since the shaded area is common to both) it will be impossible for the shaded area to be as large as the white area, no matter where we place the centre of the right-hand circle. But the left-hand circle has radius 1, so the right-hand circle must certainly have radius between $\sqrt{2}/2$ and $\sqrt{2}$. Thus it makes sense for the range for a to be 0.71 to 1.41, and for d to be 0 to 1. Now use `implicitplot`:

```
>   plots[implicitplot](eq,a=0.71..1.41,
        d=0..1);
```

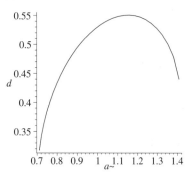

The most striking feature is that there is a single radius for which the separation d is a maximum. Maple can find this radius to be at $a = 1.154\,289\,694$ using `fsolve(eq,diff(eq,a))`.

change to `fsolve({eq, diff(eq,a)})`.

Solution 2.34

(a) To make the following code more general, we define f to be the expression in the question and define m and n to be the orders of the numerator and denominator, respectively:

```
>   f := tan(x):   m := 3:   n := 3:
```

Now we let p be a general polynomial of degree m, and let q be the general polynomial of degree n with constant term 1:

```
>   p := add(a||i*x^i,i=0..m);
    q := 1+add(b||i*x^i,i=1..n);
```

$$p := a0 + a1\,x + a2\,x^2 + a3\,x^3$$
$$q := 1 + b1\,x + b2\,x^2 + b3\,x^3$$

(b) Now use `series` to find the Taylor series of the expression $fq - p$:

```
>   fs := series(f*q-p,x,m+n+1);
```

$$fs := -a0 + (1 - a1)\,x + (-a2 + b1)\,x^2$$
$$+ \left(-a3 + b2 + \frac{1}{3}\right)x^3 + \left(b3 + \frac{b1}{3}\right)x^4$$
$$+ \left(\frac{b2}{3} + \frac{2}{15}\right)x^5 + \left(\frac{b3}{3} + \frac{2\,b1}{15}\right)x^6 + \mathrm{O}(x^7)$$

(c) Now solve the equations for the unknowns:

```
>   eq := {seq(coeff(fs,x,i),i=0..m+n)}:
>   sol := solve(eq);
```

$$sol := \{b1 = 0,\ b3 = 0,\ a2 = 0,\ a1 = 1,$$
$$a3 = \frac{-1}{15},\ b2 = \frac{-2}{5},\ a0 = 0\}$$

Now substitute this solution into the general polynomials p and q to arrive at the solution:

```
>   r := eval(p/q,sol);
```

$$r := \frac{x - \dfrac{1}{15}\,x^3}{1 - \dfrac{2\,x^2}{5}}$$

(d) Now we plot the resulting approximation on the same graph as the function:

```
>   plot([f,r],x=-2..2,-20..20,
        discont=true);
```

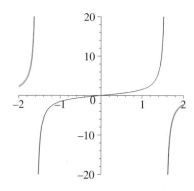

The approximation is almost indistinguishable from the original function on this interval.

UNIT 3 Programming

Study guide

This unit builds on the results of *Unit 1* and *Unit 2*, and the three sections should be studied in numerical order. Apart from reading the text and understanding the examples, it is crucial that you work through the exercises, as solving these is essential to learning the material.

You will require access to a computer to study this unit.

Introduction

In this unit we explore programming in Maple. A computer program is a collection of instructions (sometimes called an algorithm), that describe a task and enable it to be carried out by a computer. Without programs, computers are completely useless. All of the software on your PC, including the operating system, is a computer program, and almost every modern electronic device, from washing machines to aircraft, is controlled by such programs.

Computer programs are written in a 'computer programming language', of which there are thousands: C, C++, BASIC, FORTRAN and Maple are just a few. Maple is a very specialized programming language, important for scientific and mathematical applications. It and similar computer algebra languages are now routinely used in mathematical and scientific research and education.

In the previous two units, you learned how to use a number of Maple commands and how to combine them to solve a variety of mathematical problems. This approach is particularly useful when the mathematical problems are relatively simple, or for exploring different approaches to solving unfamiliar problems. However, some tasks require the use of relatively few commands very many times. When it becomes excessively laborious to input all of these commands by hand, it is necessary to have some method of automating their execution. This is achieved by writing a computer program.

There are three key programming constructs that Maple provides which are common to most programming languages, namely, loops, conditionals and procedures.

Loops enable the execution of many lines of Maple code repeatedly and sequentially. Simple loops are explored in Section 3.1, where several examples are presented in order to illustrate the sort of problems to which they are applicable. The section ends with an application of loops to explore an unproven mathematical conjecture: that the digits in the decimal representation of π occur on average with equal frequency.

In Section 3.2 we study *conditionals*. These allow different actions to be taken depending on the validity of some statement, so they allow a program to 'make decisions'. They are particularly powerful when combined with loops; this combination is discussed in Subsection 3.2.2. The section ends with a brief application of conditionals to a very famous (unproven) mathematical conjecture: Goldbach's conjecture, that every even number $N > 2$ is the sum of two primes.

The last programming construct to be covered is the *procedure*. Procedures are a way of dividing a program into re-usable chunks, so that sections of code can be used in several places and possibly several programs. Their main purpose is to simplify programming, so that a complicated task can be built up from simpler sub-tasks. In Section 3.3 we explore various aspects of procedures, and apply them to a variety of problems to illustrate their use.

The ideas developed in this unit will be used extensively in *Unit 4* and throughout the rest of the course, so it is essential that you understand them and are at ease using them. For this reason there are a relatively large number of exercises in this unit, and it is very important that you attempt these, as they are designed to help you learn and assimilate the material.

3.1   Loops

The first programming construct we examine is the **for loop**, which has analogues in many programming languages. The for loop allows one or more Maple statements to be executed repeatedly. In some respects it is similar to, and a generalization of, the `seq` command introduced in *Unit 2*.

This section introduces the loop construct, examines some typical examples of how loops are used, and finishes by applying loops to a case study: the distribution of digits in π.

3.1.1   Elementary loops

Here is an elementary example of a loop:

```
>   for m from 3 to 7 by 1 do          #1
>     m^2;                             #2
>   end do;                           #3
                    9
                    16
                    25
                    36
                    49
```

Line #1 tells Maple how many times to execute line #2, and line #3 (`end do`) simply marks the end of the loop. All loops must terminate with `end do`, although Maple also recognizes the equivalent loop terminator `od`.   od is do written in reverse.

In this case, line #1 instructs Maple to execute line #2 five times, with m assigned the values 3, 4, 5, 6 and 7 sequentially.

Warning

Notice that pressing Enter after the first line of a loop will cause Maple to issue a warning:

```
>  for m from 3 to 7 by 1 do
>
Warning, premature end of input
```

This warning will continue to be shown until you terminate the loop with **end do** (or **od**).

The warning has nothing to do with the fact that no terminator is placed at the end of line #1. You can, if you wish, end line #1 with a terminator and notice that the warning persists. As a matter of convention, we choose not to place a terminator at the end of the first line of a loop, because the loop is not finished until after the **end do**.

This warning can be suppressed by using Shift–Enter instead of Enter (see the *Computing Guide* for more information).

Now let us modify the first line of this code slightly:

```
>  for m from 3 to 7 by 2 do
>    m^2;
>  end do;
```

$$9$$
$$25$$
$$49$$

This time the variable m (called the **index**) increases by 2 on each iteration of the loop. So it takes the value $m = 3$ on the first iteration, $m = 5$ on the second, and $m = 7$ on the third.

Exercise 3.1

Predict what will happen if you change the first line to:

(a) `for m from 3 to 7 by 3 do`   (b) `for m from 3 to 7 by 5 do`

Check your predictions using Maple.

Note that the numbers occurring in the first line do not have to be integers. For example, to output the value of $\sin x$ for $x = -\frac{1}{2}, 0, \frac{1}{2}, 1$, you can use the loop

```
>  for x from -1/2 to 1 by 1/2 do
>    x,sin(x);
>  end do;
```

$$\frac{-1}{2}, -\sin\left(\frac{1}{2}\right)$$
$$0, 0$$
$$\frac{1}{2}, \sin\left(\frac{1}{2}\right)$$
$$1, \sin(1)$$

Now consider another example, which calculates $\ln(p_k)$ for $k = 1, 2, 3,$ where p_k is the kth prime number:

```
>  for k from 1 to 3 by 1 do                    #1
>     pk := ithprime(k);                         #2
>     evalf[6](ln(pk));                          #3
>  end do;                                       #4
```

Recall that `ithprime(k)` gives the kth prime number, for k a positive integer.

$$pk := 2$$
$$0.693147$$
$$pk := 3$$
$$1.09861$$
$$pk := 5$$
$$1.60944$$

Note that lines #2 and #3 (the **body** of the loop) both produce output. Now replace the noisy terminator (;) on line #4 with the silent terminator (:) and observe the effect. In general, a loop produces no output if the terminator after the **end do** is silent, but produces output from each line of its body if the terminator is noisy. The terminators in the body of the loop have no effect on the output produced.

Loops that end in a noisy terminator are called **noisy**, and loops that end in a silent terminator are called **silent**.

Often it is desirable to follow the evolution of only one or two variables within a loop whose body is many lines in length. The **print** command allows you to output specific values within a silent loop:

```
>  for k from 1 to 3 by 1 do                    #1
>     pk := ithprime(k);                         #2
>     print(pk,evalf[6](ln(pk)));                #3
>  end do:                                       #4
```

$$2, 0.693147$$
$$3, 1.09861$$
$$5, 1.60944$$

Here the loop ends with a silent terminator and we obtain output only from line #3, i.e. the values of `pk` and `log(pk)`, expressed to a convenient number of significant figures. Note that the argument of the **print** command is an expression sequence, and this command always produces output, irrespective of whether it is terminated by a noisy or silent terminator.

Exercise 3.2

Without using Maple, say what output you expect from the following loops.

These loops are input all on one line, which is perfectly acceptable.

(a) `for k from 1 to 5 by 1 do k-1:   end do;`

(b) `for k from 2 to 5 by 2 do sin(k):   end do;`

(c) `for k from 10 to 2 by -3 do k^2:   print(k,k^2):   end do:`

Check your answers using Maple, and check the value of the index `k` after these loops have executed. What do you conclude?

[*Hint*: Note that the loop terminates after the **end do**.]

Exercise 3.3

Use loops to evaluate the following numerically.

(a) $\displaystyle\sum_{k=1}^{100}\frac{(-1)^k}{k}$ (b) $\displaystyle\sum_{k=5}^{20}\frac{1}{1-k^2}$ (c) $\displaystyle\prod_{k=2}^{10}\frac{1}{1+\sin k}$

(Note that in practice it would be better to use the Maple commands `add` and `mul` to evaluate such simple sums and products.)

From the preceding examples, observe that loops have the form:

```
>   for var from n1 to n2 by n3 do
>       body of
>           loop
            ⋮
>   end do;
```

Of course, the loop could also end with `end do:`.

where *var* is a variable, and *n1*, *n2* and *n3* are expressions which evaluate to numbers, and the **body** of the loop is one or more Maple statements.

The variable *var* is called the **index** of the loop, *n1* and *n2* are the **lower** and **upper bound**, respectively, and *n3* is the **increment** or **step** of the loop.

For example, the following loop evaluates the function $\sin x$ at the points $x = \pi,\ \pi + 0.3,\ \pi + 0.6$:

```
>   for x from evalf(Pi) to 4 by 0.3 do
>       x,sin(x):
>   end do;
```

$$3.141592654,\ -0.4102067615\ 10^{-9}$$
$$3.441592654,\ -0.2955202071$$
$$3.741592654,\ -0.5646424737$$

Here the index is `x`, the step is `0.3`, the lower bound is $\pi \simeq 3.141\,592\,654$, and the upper bound is `4`.

You may have noticed that we indent the body of the loop using the spacebar. This is not necessary, but it is good practice, as it helps the reader to distinguish it from lines of Maple outside the loop. (Pressing the Tab key will *not* produce an indent in Maple input.)

This example also illustrates a common potential source of error. Observe what happens if you replace the first line of the loop with the superficially similar

```
>   for x from Pi to 4 by 0.3 do
```

Maple outputs an error message and the loop is not executed. The reason for this is that the bounds and the step have to evaluate to numbers, and in this case the lower bound `Pi` is unevaluated. Another common source of error occurs when the index is assigned values within the loop. This should be avoided at all costs, as it may stop the loop behaving as expected.

If you don't explicitly give values for the bounds and the step, then they take their default values. The default value of the lower bound *n1* and the increment *n3* is 1, while the default value of the upper bound *n2* is `infinity`. So, for example, `for k from 1 to infinity by 1 do` is equivalent to `for k do`.

In Subsection 3.2.1 we will present a construct which will enable us to terminate the execution of a loop whose upper bound is `infinity`.

Exercise 3.4

What are the non-default forms of the following?

(a) `for k from -3 to 7 do`

(b) `for k to 7 by 3 do`

(c) `for k from -1 by 2 do`

Warning

Once the entire loop has been input, observe that it is enclosed in the worksheet by a single square bracket on the left, indicating that Maple regards it as a single executable statement. Sometimes editing a loop can lead to the single executable statement being broken, so that instead of there being just a single square bracket on the left, there may be two or more. For example:

```
>   for k from 1 to 3 by 1 do
>      pk := ithprime(k);

>      print(pk,evalf[6](log(pk)));
>   end do:
```

To remedy this, place the cursor on the first part of the loop, and press the **F4** key to join the two execution groups together.

If you place the cursor at the end of the second line of the print example on page 110 and use the **F3** key, you will break the loop as shown here.

3.1.2   Some examples of loops in action

Let us now look at some examples which illustrate various uses of loops.

Example 3.1

Use a loop to evaluate x_1, x_2, x_3, x_4, both as rationals and as decimal approximations, where

$$x_{k+1} = \frac{x_k}{2} + \frac{5}{2x_k} \quad \text{with} \quad x_0 = 1 \quad \text{and} \quad k = 0, 1, 2, \ldots . \tag{3.1}$$

In the limit as $k \to \infty$, the sequence converges to $\sqrt{5}$. You may like to show this using the fact that, in this case, $x_{k+1} \to x_k$ as $k \to \infty$.

Solution

```
>   x := 1:                          # initial value
>   for k to 4 do
>      x := x/2+5/(2*x):             # iterative formula
>      print(k,x,evalf(x)):          # output result
>   end do:
```

$$1, 3, 3.$$

$$2, \frac{7}{3}, 2.333333333$$

$$3, \frac{47}{21}, 2.238095238$$

$$4, \frac{2207}{987}, 2.236068896$$

Here we have used a trick whereby the Maple variable x consecutively stands for $x_0, x_1, \ldots, x_4$.

On the first line we assign the value 1 to x, corresponding to the initial condition $x_0 = 1$.

So on the first iteration of the loop (where $k = 1$), x is assigned the new value $1/2 + 5/2 = 3$, which corresponds to $x_1 = x_0/2 + 5/(2x_0)$.

On the second iteration (where $k = 2$), x is assigned the value $3/2 + 5/6$, corresponding to x_2.

And so on until $k = 4$.   ♦

Loops are often used to build expression sequences. To do this it is useful to initialize the sequence with the NULL sequence, which is simply the sequence with no elements. The following example illustrates the method.

If you have studied M203 or M208, please note that the concept of 'null sequence' studied there has nothing to do with the NULL sequence defined here.

Example 3.2

Observe how the sequence x_1, x_2, x_3, x_4 from the previous example approaches $\sqrt{5}$, by plotting $\log_{10}|x_k - \sqrt{5}|$ versus k for $k = 1, 2, 3, 4$.

Solution

We need to plot the logarithm of the difference $|x_k - \sqrt{5}|$ because this quantity approaches zero very quickly.

Recall from *Unit 2* that in order to plot a set of coordinates $(x_1, y_1), (x_2, y_2), (x_3, y_3), \ldots$, it is necessary to produce a list of the form

> ` dat := [[x_1, y_1], [x_2, y_2], [x_3, y_3], ...];`

More precisely, this is a list of lists.

which is plotted using the command `plot(dat)`.

In this case we require the list

$$\Big[[1, \log_{10}|x_1 - \sqrt{5}|], \ [2, \log_{10}|x_2 - \sqrt{5}|],$$
$$[3, \log_{10}|x_3 - \sqrt{5}|], \ [4, \log_{10}|x_4 - \sqrt{5}|] \Big],$$

which can be obtained from the expression sequence Xs produced by the following loop:

```
>   x := 1:                                            #1
>   Xs := NULL:              # initialize the sequence Xs   #2
>   for k to 4 do                                      #3
>     x := x/2+5/(2*x):                                #4
>     Xs := Xs,[k,log[10](abs(x-sqrt(5)))]:            #5
>   end do:                                            #6
>   evalf[6](Xs);                                      #7
        [1., -0.116945], [2., -1.01204], [3., -2.69308], [4., -6.03668]
>   dat := [Xs]:  plot(dat);                           #8
```

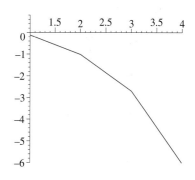

Figure 3.1

The plot is displayed in Figure 3.1. Clearly, x_k converges very quickly to $\sqrt{5}$ (with each iteration adding more significant figures to the accuracy than the previous one), so $|x_4 - \sqrt{5}| \simeq 10^{-6}$.

The code is similar to that of Example 3.1, but with the addition of lines #2 and #5.

On lines #1 and #2, x is initialized to 1, and Xs (an expression sequence) is initialized to the NULL sequence.

On the first iteration of the loop (where $k = 1$), x is assigned the value 3 and Xs the value $[1, \log_{10}|3 - \sqrt{5}|]$.

113

On the second iteration of the loop (where $k = 2$), x is assigned the value $7/3$ and Xs the value $[1, \log_{10}|3 - \sqrt{5}|], [2, \log_{10}|7/3 - \sqrt{5}|]$.

The iterations continue until the complete expression sequence is built up, evaluated and displayed following line #7, then converted to a list (which we have named dat) and plotted on line #8.

This technique of building up an expression sequence by starting with NULL and appending elements is one which will occur frequently.   ◆

Note that it is not always necessary to start with NULL. For example, in this case, we could have assigned Xs to the first element in line #2, and then looped over $k = 2, 3, 4$ to achieve the same result.

In *Unit 2* you used repeated composition to calculate F_{100}.

Example 3.3

The *Fibonacci sequence* $F_1, F_2, F_3, \ldots$ is an infinite sequence of numbers defined by the formulae

$$F_1 = 1, \quad F_2 = 1, \quad F_n = F_{n-2} + F_{n-1} \quad \text{for} \quad n \geq 3.$$

So it begins $1, 1, 2, 3, 5, \ldots$.

Use a loop to generate the first six elements of the sequence.

Solution

```
> F1 := 1;  F2 := 1;                                          #1
```
$$F1 := 1$$
$$F2 := 1$$
```
> for k from 3 to 6 do                                        #2
> F := F1+F2:                                                 #3
> F1 := F2:                                                   #4
> F2 := F:                                                    #5
> print(F||k=F):                                              #6
> end do:
```
$$F3 = 2$$
$$F4 = 3$$
$$F5 = 5$$
$$F6 = 8$$

This code contains three variables, F1, F2 and F. On line #1, the sequence is initialized. The loop calculates the remaining F_k, assigns them to the variable F, and outputs the result on line #6.

In order to understand how the loop works, consider the values of the variables F1, F2, and F on line #6 for each iteration of the loop.

Initially, on entering the loop, we have the values

F1 = 1, F2 = 1.

Then for successive iterations of the loop we have the values

k = 3, F = 2, F1 = 1, F2 = 2,
k = 4, F = 3, F1 = 2, F2 = 3,

and so on.

Note that the print command, on line #6, uses concatenation (see *Unit A2*) to make the output more elegant. The first two lines of output therefore differ slightly in appearance from the rest.   ◆

The syntax for the argument of print is not very intuitive, and you may have to refer back to this example when you use this command.

Example 3.4

Use a loop to calculate the 5th-order Taylor series expansion of $\tan x$ about $x = 0$.

(Normally you would use the command **series** to do this, but using a loop is an instructive exercise.)

Solution

The 5th-order Taylor series expansion of a function $f(x)$ about $x = 0$ is given by

$$f(x) = \sum_{n=0}^{5} c_n x^n \quad \text{where} \quad c_n = \frac{f^{(n)}(0)}{n!}.$$

To calculate the 5th-order expansion of $\tan(x)$ we proceed as follows:

```
>   restart:  f := tan:                                        #1
>   c0 := f(0):            # define c0                          #2
>   ftay := c0:            # 0th-order Taylor expansion         #3
>   for n to 5 do                                              #4
>       f := D(f):                                             #5
>       cn := f(0)/n!:                                         #6
>       ftay := ftay+cn*x^n:                                   #7
>   end do:                                                     #8
>   ftay;
```

Recall from *Unit 2* that D is the differential operator.

$$x + \frac{1}{3}x^3 + \frac{2}{15}x^5$$

which is the same result as obtained using the **series** command, but as a polynomial (i.e. without the O term).

On lines #1, #2 and #3 we define the function f, the coefficient $c_0 = f(0) = 0$, and the 0th-order Taylor expansion c_0, which we assign to the variable **ftay**.

On the first iteration of the loop, at line #7, the variables **n**, **f**, **cn** and **ftay** have the values

$$\mathbf{n} = 1, \ \mathbf{f} = f^{(1)}(x), \ \mathbf{cn} = f^{(1)}(0)/1!, \ \mathbf{ftay} = c_0 + x f^{(1)}(0)/1!.$$

On the second iteration,

$$\mathbf{n} = 2, \ \mathbf{f} = f^{(2)}(x), \ \mathbf{cn} = f^{(2)}(0)/2!, \ \mathbf{ftay} = c_0 + x f^{(1)}(0)/1! + x^2 f^{(2)}(0)/1!,$$

and so on.   ◆

Exercise 3.5

Use loops to calculate the following.

(a) $10! = 10 \times 9 \times 8 \times \cdots \times 2 \times 1$

(b) $20 \times 17 \times 14 \times \cdots \times 5 \times 2$

Exercise 3.6

Consider the iterative sequence defined by $x_{k+1} = \frac{1}{n}\left(\frac{N}{x_k^{n-1}} + (n-1)x_k\right)$,

for $k = 0, 1, 2, \ldots$, beginning with $x_0 = 1$. It can be shown that as $k \to \infty$, this sequence converges to $N^{1/n}$. (In fact, this sequence can be obtained by using the Newton–Raphson method to solve $f(x) = 0$ where $f(x) = x^n - N$.)

> This formula was used to generate the approximation to $\sqrt{5}$ in Example 3.1 for $N = 5$, $n = 2$.

Specify the value of k that is required for x_k to be accurate to 8 decimal places for:

(a) $\sqrt{3}$ (b) $7^{1/4}$ (c) $23^{1/3}$.

Exercise 3.7

Consider the rational function $f_n(x)$, for $n = 1, 2, 3, \ldots$, which is defined by the sequence $f_{n+1}(x) = (1 - xf_n(x))^{-1}$, beginning with the function $f_1(x) = 1$.

> Recall that a rational function is one of the form $\frac{P(x)}{Q(x)}$, where $P(x)$ and $Q(x)$ are polynomials.

(a) Use a loop to show that $f_8(x) = -\dfrac{-1 + 6x - 10x^2 + 4x^3}{1 - 7x + 15x^2 - 10x^3 + x^4}$.

(b) $f_\infty(x)$ is obtained by solving the equation $f_\infty(x) = (1 - xf_\infty(x))^{-1}$. Solve this equation for the two solutions $f_\infty(x)$, and plot them together with $f_8(x)$ for the values $-20 \le x \le 0$, $-1 \le y \le 1$. Note that since $f_n(0) = 1$ for all n, only one solution for $f_\infty(x)$ is correct.

We will return to loops in Subsection 3.2.2, after we introduce conditionals.

3.1.3 Case study: The distribution of digits in π

It is generally believed that as π is expressed to more and more decimal places, each digit will occur on average 1/10th of the time, but this conjecture has never been proved.

For example, π expressed to 5 significant figures is 3.1415, which has no occurrences of the digits 0, 2, 6, 7, 8, 9, one occurrence of each of the digits 3, 4, 5, and two occurrences of the digit 1. In Figure 3.2 we display a graph of the frequency with which each digit occurs. (The commands to construct this graph are presented on page 119.)

> Note that we are not rounding up the last digit here.

In this subsection, we will use Maple to obtain similar graphs for π expressed to much larger numbers of significant figures. In this way we can at least check 'experimentally' that each digit occurs with approximately the same frequency.

The problem naturally splits into three elementary parts:

(1) it is necessary to express π to a given number of significant figures and to extract each digit sequentially;

(2) the number of occurrences of each digit must be calculated;

(3) these frequencies must be manipulated into a convenient form for display.

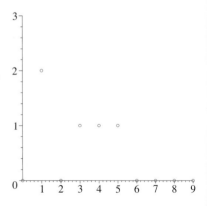

Figure 3.2 Frequency of digits in 3.1415

Extracting each digit from π

There are many methods to extract the digits of π. Here we present a conceptually simple algorithm involving just two commands, `floor` and `frac`. Alternative methods will be set as exercises.

To begin with, suppose that we just want to extract the first 5 digits of π. Then we assign a variable N:=5, and calculate π to six significant figures, in order to avoid the possibility of the last digit being rounded up:

```
>  N := 5:  pi1 := evalf[N+1](Pi);
```
$$\pi1 := 3.14159$$

If x is a positive real number, the Maple command `floor(x)` gives the integer part of x, and `frac(x)` gives the fractional part of x:

```
>  dig := floor(pi1);
```
$$dig := 3$$

```
>  frac(pi1);
```
$$0.14159$$

The command `floor(pi1)` therefore gives the first digit in the decimal expansion of π. Now notice that if the fractional part of `pi1` is multiplied by 10, we obtain 1.4159, so the integer part of this is the second digit of π:

```
>  pi1 := evalf[N+1](10*frac(pi1));
```
$$\pi1 := 1.4159$$

```
>  dig := floor(pi1);
```
$$dig := 1$$

Since `pi1` now has the value 1.4159, the third digit of π can be obtained using the same method, i.e. multiply the fractional part of `pi1` by 10, re-assign the result to the variable `pi1`, and take the integer part:

```
>  pi1 := evalf[N+1](10*frac(pi1));
```
$$\pi1 := 4.159$$

```
>  dig := floor(pi1);
```
$$dig := 4$$

Proceeding in this manner, all the digits in `pi1` can be extracted. However, we would ultimately like to examine perhaps thousands of digits, so we use a loop to automate the process:

Code 3.1

```
>  restart:  N := 5:  pi1 := evalf[N+1](Pi):        #1
>  for k from 1 to N do                              #2
>     dig := floor(pi1):                              #3
>     pi1 := evalf[N+1](10*frac(pi1)):               #4
>  end do;                                            #5
```

Note that because line #1 initializes the loop variables, it is good practice to make all these lines into a single execution group.

You should examine the output from this code, and observe that the loop extracts each digit of π in turn, and assigns it to the variable `dig`.

Exercise 3.8

Adapt the above method to find a neat way of extracting the 10 001st to 10 005th digits of π.

Exercise 3.9

Code 3.1 extracts the digits of π using `frac`, `floor` and `evalf`. Create equivalent code which achieves the same results using only `floor` and `evalf`.

Exercise 3.10

Create a loop to extract the 3-digit sequences $314, 141, 415, \ldots$ from π.

(The results of this exercise are used later, in Exercise 3.36.)

Counting the digits of π

Now that the digits of π have been extracted automatically, a method of counting the number of 0s, 1s, 2s, etc. is required.

The way we do this is to define a list **num** with 10 elements. The value of the first element, `num[1]`, will correspond to the number of 0s in the decimal expansion of π. The second element, `num[2]`, will give the number of 1s, the third element, `num[3]`, the number of 2s, etc.

The following code calculates **num**. It is the code above, with the addition of two extra lines, here labelled #2 and #6.

Code 3.2

```
>   restart:   N := 5:   pi1 := evalf[N+1](Pi):        #1
>   num := [0$10]:                                      #2
>   for k from 1 to N do                                #3
>     dig := floor(pi1):                                #4
>     pi1 := evalf[N+1](10*frac(pi1)):                  #5
>     num[dig+1] := num[dig+1]+1:                       #6
>   end do:                                             #7
```

Recall that $10 means 'repeated ten times'.

As this new code is executed, the values of `dig` and `pi1` will not differ from those found by Code 3.1, so at the kth iteration of the loop, `dig` is assigned to the kth digit of π.

The additional lines simply calculate the elements of **num**. Line #2 initializes **num** to `[0,0,0,0,0,0,0,0,0,0]`, indicating that at the beginning, no digits in the expansion of π have been counted. Line #6 alters the values of its elements. Let us examine how.

On the first iteration of the loop $(k = 1)$, the values `dig:=3` and `pi1:=1.41590` will be assigned. Hence line #6 becomes `num[4]:=0+1`, giving **num** the value `[0,0,0,1,0,0,0,0,0,0]`, corresponding to the fact that the first digit in the decimal expansion of π is 3.

The second time line #6 is executed, the values `dig:=1` and `pi1:=4.15900` have been assigned. Hence line #6 becomes `num[2]:=0+1`. So the value `num := [0,1,0,1,0,0,0,0,0,0]` will be assigned at the end of the second iteration.

The loop continues for five iterations, at the end of which **num** has the value

```
>   num;
```
$$[0, 2, 0, 1, 1, 1, 0, 0, 0, 0]$$

which shows that in the expansion 3.1415, there are no 0s, 2s, 6s, 7s, 8s or 9s, there is one occurrence each of 3, 4 and 5, and there are two occurrences of 1.

118

All this is rather obvious for such a small expansion of π. However, the beauty of this code is that it will work when π is expressed to thousands of significant figures, simply by altering the value of the variable N on the first line. Note, however, that for large values of N it is wise to end the loop with the silent terminator.

Exercise 3.11

Show that there are 105 occurrences of 9 in the first 1000 digits of π.

Exercise 3.12

Show that there are 90 occurrences of 1 between the 2000th and 3000th digits (inclusive) of the exponential number e.

Plotting the frequency of digits of π

In order to plot the list `num := [0,2,0,1,1,1,0,0,0,0]`, it is necessary to create a data list of the form

   `[[0,0],[1,2],[2,0],[3,1],[4,1],[5,1],[6,0],[7,0],[8,0],[9,0]]`.

More generally, we require a list `num` of the form

   `[[0,num[1]],[1,num[2]],...,[9,num[10]]]`.

One way to do this is to use the `seq` command:

```
>  dat := [seq([k-1,num[k]],k=1..10)];
```

which can be plotted using `plot(dat);`.

Instead of plotting n_i, the number of times the digit i occurs, it is preferable to plot the relative frequencies n_i/N, which can be done by modifying `dat` to

```
>  dat := [seq([k-1,num[k]/N],k=1..10)];
```

To reproduce Figure 3.2 on page 116, you need to include the following options:

```
plot(dat,view=[0..9,0..3],
style=point,symbol=circle,
symbolsize=15,
tickmarks=[9,3]);
```

In Figures 3.3–3.5, we plot the relative frequencies of each digit in the decimal expansion of π to $N = 100$, $N = 1000$ and $N = 10\,000$ significant figures, respectively. Observe from these plots the trend that, as N increases, the relative frequencies of the digits become approximately equal, i.e. $n_i/N \simeq 1/10$, $i = 0, 1, \ldots, 9$.

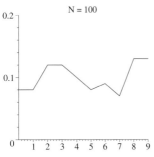

Figure 3.3

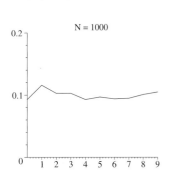

Figure 3.4

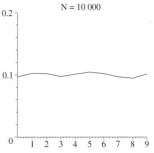

Figure 3.5

Exercise 3.13

One way to think about the decimal representation of π is to write

$$\pi = 3.1415\ldots = 3 \times 10^0 + \frac{1}{10^1} + \frac{4}{10^2} + \frac{1}{10^3} + \frac{5}{10^4} + \cdots.$$

Similarly, the hexadecimal representation of π (i.e. base 16) has the form

$$\pi = c_0 \times 16^0 + \frac{c_1}{16^1} + \frac{c_2}{16^2} + \frac{c_3}{16^3} + \cdots,$$

where the hexadecimal digits of π are the integers $c_0, c_1, \ldots$, which all have values between 0 and 15.

(a) Modify Code 3.1 to calculate the first 5 hexadecimal digits of π $(c_0, \ldots, c_4)$.

[*Hint*: You will have to represent π to at least `floor(log[10](16^5))+1` decimal significant figures to do this.]

(b) Modify Code 3.2 to determine the relative frequency of each hexadecimal digit for the first 10 000 digits, and display your results on a plot similar to Figure 3.5.

3.2 Conditional statements

The next programming tool covered in this unit is the **conditional statement**. It is often necessary to take different actions depending on whether a statement is true or false. The conditional statement (also called the **if-conditional** or just **conditional**) is designed for this purpose, and allows a program to 'make decisions'. It is common to most programming languages, and in Maple it can take many different forms.

This section begins with some elementary examples of conditionals, moves on to show how conditionals can interact with loops, and finishes by applying loops to a case study: *Goldbach's conjecture*.

3.2.1 Basic conditionals

Suppose that two variables aa and bb are assigned the value 3:

```
>  aa := 3:  bb := 3:
```

Then executing the conditional statement

```
>  if aa=bb then print("aa and bb are the same.") end if:
```

produces the result

> "aa and bb are the same."

Altering the assignments on the first line, so that the values of aa and bb are not the same, produces no output when the conditional is executed. So this conditional produces the output "aa and bb are the same." if the value of aa is the same as bb, but produces no output if their values differ.

Now consider another example:

```
> aa := 3:  bb := 4:
> if aa<>bb then print("aa and bb are not the same.") end if:
```

In this case the conditional statement produces the output
"aa and bb are not the same." if the value of **aa** differs from **bb**, but
produces no output if their values are the same.

The operator **<>** here is the inequality operator and is analogous to the $\neq$
sign used in mathematics. This and the other **comparison operators**
were introduced in *Unit 2*, in Table 2.1.

A less trivial use of the conditional statement is provided in the following
example, where we define a function f which takes integer arguments i.
(Note that, as with loops, we choose to indent lines of code contained
within conditionals in order to make the text more readable.)

```
> f := i->
>     if ithprime(i)>10^5 then
>       print("the "||i||"th prime number is greater
              than 10^5"):
>       print(ithprime(i)):
>     end if:
```

When you begin to input this
function you will receive an
error message after each line
(unless you end each line with
Shift–Enter). The message
will persist until you type
end if:. This function and
all conditionals in general
must be in a single execution
group.

If the ith prime number p_i is greater than 10^5, then the string
"the ith prime number is greater than 10^5" and the value of p_i are
output. No output is produced if p_i is less than 10^5. For example:

```
> f(10000);
```
         "the 10000th prime number is greater than 10^5"
                          104729

The basic form of the if-conditional is

```
> if conditional expression then action sequence end if;
```

Note that the conditional begins with an **if** and ends with an **end if**.
Alternatively, it can end with **fi** (which is **if** written in reverse). The
conditional expression is any expression which Maple can check to be
true or false, whereas the **action sequence** is one or more Maple
statements. When Maple encounters a conditional, it executes the action
sequence if and only if the conditional expression evaluates to true.

The conditional expression can be formed by using the comparison
operators together with the **logical operators and, or** and **not**.

In the following example, we use a conditional to check, for two integers x
and y, whether one is divisible by the other. This is true if x and y are
non-zero (i.e. $xy \neq 0$) and either x/y or y/x is an integer.

```
> f := (x,y)->
>    if x*y<>0 and (frac(x/y)=0 or frac(y/x)=0) then
>      print("One of "||x||" and "||y||" is divisible by the
              other."):
>    end if:
> f(3,9);
```
          "One of 3 and 9 is divisible by the other."

Recall that the **frac** command gives the fractional part of a number. So if
frac(M/N)=0 then **M** is divisible by **N**. Note that the outer brackets in the
above conditional are necessary for it to be unambiguous.

The clause 'if A and B or C,
then ...' can be interpreted
in two ways, one of which
allows just C to be true and
the other of which does not.

In addition to the comparison operators, Maple provides a variety of built-in functions whose values are either *true* or *false*. You have already met one such function, isprime, and learned how to define your own in *Unit 2*. These can also be used in conditional statements.

Conditionals are very often used in conjunction with loops, and Maple provides two associated commands, next and break, which are particularly useful. The next command forces a loop to abandon its current iteration and to proceed to the next one. The following example creates an expression sequence Xs whose elements are the non-prime numbers between 1 and 10.

```
>  Xs := NULL:
>  for k to 10 do
>     if isprime(k) then next end if:
>     Xs := Xs,k:
>  end do:
>  Xs;
```

$$1, 4, 6, 8, 9, 10$$

The break command, which is illustrated in the following example, terminates the iterations of a loop prematurely (that is, before the upper bound of the index has been reached).

Example 3.5

Construct a loop which tests if an integer N is a factorial; that is, if $N = m!$ for some m, then the loop finds and prints m.

Solution

```
>  N := 6227020800:
>  for k do                                     #1
>     if k!=N then print(k) end if:             #2
>     if k!>N then break end if:                #3
>  end do:                                      #4
```

$$13$$

The loop iterates over $k = 1, 2, \ldots$. Line #2 prints N if $k! = N$. Line #3 terminates the execution of the loop when $k! > N$. No output is produced if N is not a factorial. Here the loop has verified that $6\,227\,020\,800 = 13!$. ◆

Exercise 3.14

Use a loop to create an expression sequence whose elements are the integers between 1 and 50 which are neither prime, nor even, nor divisible by 3.

Exercise 3.15

Use a loop to produce all the prime numbers greater than 5000 but less than 5050. Display the results as an expression sequence.

Exercise 3.16

(a) Show that the formula $n^2 - n + 41$ gives prime numbers for $n = 0, 1, \ldots, 40$.

(b) What is the smallest integer value of n for which $n^2 - 79n + 1601$ is not prime?

---

else *and* elif *options*

The if syntax can be extended by use of the else option. This is particularly useful when it is necessary to distinguish between two alternatives. For example, the following function classifies integers as even or odd.

```
>   f := N->
>       if frac(N/2)=0 then print(""||N||" is even."):
>       else print(""||N||" is odd."):
>       end if:
>   f(5);
```

"5 is odd."

If a is a string, then a||b is also a string provided that b is an integer or a variable. So since "" is a null string, ""||N|| is a convenient way of converting N to a string.

Here the print(""||N||" is odd.") statement is executed only if $N/2$ has a fractional part. This conditional therefore has the form

```
>   if condition1 then action1:
>   else action2:
>   end if:
```

If *condition1* is true, then *action1* is executed, otherwise *action2* is executed.

The most general form of conditional is obtained by using the else and elif options. In this case the general form of the conditional is

elif is an abbreviation for *else if*.

```
>   if condition1 then action1:
>   elif condition2 then action2:
        ⋮
>   elif conditionN then actionN:
>   else action:
>   end if:
```

Note that there does not have to be an else clause.

In such a conditional, each of the expressions *condition1*, *condition2*, ... is evaluated sequentially. The first expression which Maple evaluates to be *true* has its associated *action* executed, and no further conditional expressions are then evaluated. If all of the conditions are evaluated to be *false*, then Maple will execute the *action* sequence after the else, if the else clause exists.

As an example, all integers can be expressed in one of four ways: $4k$, $4k + 1$, $4k + 2$ or $4k + 3$ for some $k = 0, 1, \ldots$. The following Maple function performs this classification for any integer.

```
>  f := N->
>     if frac(N/4)=0 then print(""||N||" has the form 4k."):
>     elif frac((N-1)/4)=0 then
          print(""||N||" has the form 4k+1."):
>     elif frac((N-2)/4)=0 then
          print(""||N||" has the form 4k+2."):
>     else print(""||N||" has the form 4k+3."):
>     end if:
```

For example:

```
>  f(334);
```

$$\text{“334 has the form 4k+2.”}$$

Exercise 3.17

Use conditionals to create a function of two variables, x and y, whose output is a list $[x, y]$ if $x \geq y$ and $[y, x]$ if $y > x$.

Exercise 3.18

The solutions of a quadratic equation $ax^2 + bx + c = 0$, with real coefficients $a \neq 0$, b, c, can be classified as follows:

$b^2 - 4ac = 0$,   one real solution,
$b^2 - 4ac > 0$,   two real solutions,
$b^2 - 4ac < 0$ and $b = 0$,   two imaginary solutions,
$b^2 - 4ac < 0$ and $b \neq 0$,   two complex solutions (not pure imaginary).

Create a Maple function $f(a, b, c)$ whose output prints the classification for coefficients a, b, c.

Piecewise-defined functions

Functions defined piecewise are readily created using conditionals. For example, to create the function

$$f(x) = \begin{cases} \sin x & \text{if } \sin x \leq 1/2, \\ 1/2 & \text{if } \sin x > 1/2, \end{cases} \tag{3.2}$$

depicted in Figure 3.6 opposite, we can use the code

```
>  f := x->                                          #1
>     if evalf(sin(x))<=1/2 then sin(x):            #2
>     else 1/2:                                      #3
>     end if:                                        #4
```

Note that if you replace line #2 with if sin(x)<=0.5 then sin(x):, and attempt to evaluate $f(3)$, you will receive a warning. This is because Maple leaves $\sin 3$ unevaluated, and so cannot tell if the conditional is true.

When we try to plot this function, we obtain:

```
>  plot(f(x),x=-3*Pi..3*Pi);
```

```
Error, (in f) cannot determine if this expression is true or
false: sin(x)-.5 <= 0
```

This anomaly occurs whenever a user-defined function `f(x)` contains a conditional. It can be remedied by either of the devices that you saw in *Unit 2*, Section 2.4. The simpler of these is to place single forward quotes around the function name, which delays evaluation of `f(x)` until `x` has a numerical value:

> `plot('f'(x),x=-3*Pi..3*Pi);`

The other method is to change the command to the form

> `plot(f,-3*Pi..3*Pi);`

Either way, the output is displayed as in Figure 3.6.

Such problems can be avoided if you define functions using the `piecewise` command. For example, the function defined in Equation (3.2) can be expressed as

> `f := x->piecewise(sin(x)<1/2,sin(x),1/2);`

$$f := x \rightarrow \text{piecewise}\left(\sin(x) < \frac{1}{2}, \sin(x), \frac{1}{2}\right)$$

Functions defined in this way can be plotted as usual (the following command produces Figure 3.6):

> `plot(f(x),x=-3*Pi..3*Pi);`

They can also be differentiated:

> `diff(f(x),x);`

$$\begin{cases} \cos(x) & \sin(x) < \dfrac{1}{2} \\ 0 & \textit{otherwise} \end{cases}$$

Under some circumstances (although not in this case) they can be integrated as well.

The general form of the `piecewise` command is

$$\text{piecewise}(cond_1, f_1, cond_2, f_2, \ldots, cond_n, f_n, f_{\text{otherwise}}):$$

where $cond_1, cond_2, \ldots, cond_n$ is a set of conditional expressions, and $f_1, f_2, \ldots, f_n, f_{\text{otherwise}}$ is a set of expressions.

Its value is obtained as follows. Maple tests each of the conditions $cond_1, cond_2, \ldots, cond_n$ in turn. If $cond_i$ is the first condition found to be true, then the value is f_i. If all the conditions are false, then if it exists $f_{\text{otherwise}}$ is the value; if it doesn't exist then the value is zero by default.

As with conditionals, the conditional expressions here are made up from the six comparison operators, together with **and**, **or** and **not**. For example, the function

$$T(x) = \begin{cases} 1 & \text{if } -1 < x < \frac{1}{2}, \\ 0 & \text{otherwise} \end{cases}$$

(whose graph is shown in Figure 3.7) can be created as follows:

> `T := x->piecewise(x>-1 and x<1/2,1,0);`

$$T := x \rightarrow \text{piecewise}\left(-1 < x \text{ and } x < \frac{1}{2}, 1, 0\right)$$

This problem also occurs for functions defined using procedures (see Section 3.3).

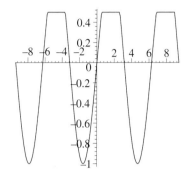

Figure 3.6   A graph of $f(x)$

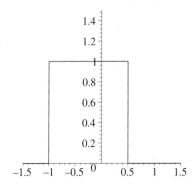

Figure 3.7   A graph of $T(x)$

125

Exercise 3.19

Use both conditionals and `piecewise` to plot the following functions over the range $-2\pi \le x \le 2\pi$.

(a) $f(x) = \begin{cases} \sin x & \text{if } |\sin x| \le 1/2 \\ 1/2 & \text{if } \sin x > 1/2 \\ -1/2 & \text{if } \sin x < -1/2 \end{cases}$

(b) $g(x) = \begin{cases} \sqrt{2x - x^2} & \text{if } 0 < x < 2 \\ -\ln(1 - x) & \text{if } x \le 0 \\ -\ln(x - 1) & \text{otherwise} \end{cases}$

3.2.2 More about loops

In Section 3.1 the basic form of the loop was introduced. In this subsection we expand on this, and briefly discuss some useful alternative versions.

The `while` option

As mentioned in Subsection 3.2.1, it is frequently necessary to combine loops with conditionals which terminate the loop prematurely using the `break` statement. Maple provides an alternative form of the loop, where the conditional terminating a loop is expressed as part of the main statement.

To illustrate this, consider Example 3.5, where a loop was created to determine if an integer N is a factorial. Compare that with the following loop, which produces exactly the same output:

```
>  N := 6227020800:
>  for k while k!<=N do                              #1
>    if k!=N then print(k) end if:                   #2
>  end do:                                           #3
```

13

This code includes a `while` statement in line #1, which has the following effect. The loop executes line #2 for $k = 1, 2, \ldots$ so long as $k! \le N$. As soon as $k! > N$, the action of the loop is terminated. Hence this loop performs an identical task to that in Example 3.5.

Another example is provided by Exercise 3.15, where you were asked to produce a list of all prime numbers greater than 5000 but less than 5050. This can be achieved with a `while` statement. It is also helpful to know of the existence of the Maple built-in function `nextprime`, which outputs the least prime greater than the input. This can be used as follows:

```
>  Xs := NULL:                                       #1
>  pk := nextprime(5000):                            #2
>  for kdum while pk<5050 do                         #3
>    Xs := Xs,pk:                                     #4
>    pk := nextprime(pk):                            #5
>  end do:                                           #6
>  [Xs];                                             #7
```

Note that the index is not used in the body of the loop. Hence we choose to call it kdum to show that it is a 'dummy variable', whose actual value is of no interest.

$$[5003, 5009, 5011, 5021, 5023, 5039]$$

Here `pk` is a prime number, which is initialized on line #2 to 5003, the first prime number greater than 5000. Each time the body of the loop is executed (lines #4 and #5), `pk` is reassigned to the next prime number, and this is appended to the expression sequence `Xs`. When `pk` exceeds 5050, the loop is terminated.

Using lists and sequences to define the index

Another way of defining the values of an index is to use a list or sequence. For example:

```
>   for k in [1,3,5,7,4,2] while k<=5 do
>       print(k):
>   end do:
```

$$1$$
$$3$$
$$5$$

In this case, as the loop iterates, the index `k` takes successive values from the list `[1,3,5,7,4,2]`. The `while` option, however, forces the loop to terminate as soon as it encounters a value greater than 5.

The technique works equally well if the square brackets are omitted from the list, so that it becomes an expression sequence.

In fact the list (or sequence) need not even be numerical, as in the following example.

```
>   func := sin,cos,tan:   sm := 0:
>   for k in func do
>       sm := sm+k(x):
>   end do:
>   sm;
```

$$\sin(x) + \cos(x) + \tan(x)$$

Nested loops

It is often necessary to have loops within other loops. Such constructs are called **nested loops**. For example, the following loop calculates the sum

$$f = \sum_{r=1}^{10} \sum_{s=4}^{20} e^{-r-2s}.$$

```
>   f := 0:
>   for r from 1 to 10 do
>       for s from 4 to 20 do
>           f := f+evalf(exp(-r-2*s)):
>       end do:
>   end do:
>   f;
```

$$0.0002257783496$$

Notice how the indentations in this code (which are not necessary for its execution) aid its interpretation.

Nested loops are frequently used to input elements of a matrix or array:

```
>  M := Matrix(3,3):
>  for i from 1 to 3 do
>    for j from 1 to 3 do
>      if i<>j then M[i,j] := x:                        #1
>      else M[i,j] := y^3:                              #2
>      end if:
>    end do:
>  end do:
>  M;
```

$$\begin{bmatrix} y^3 & x & x \\ x & y^3 & x \\ x & x & y^3 \end{bmatrix}$$

In this nested loop, lines #1 and #2 are executed for all values of the indices $i, j = 1, 2, 3$. Line #1 assigns every off-diagonal element the value x, and line #2 assigns every diagonal element the value y^3.

Exercise 3.20

Use nested loops to create the following matrices.

(a) $\begin{pmatrix} 1 & 0 & 0 \\ 1 & 1 & 0 \\ 1 & 1 & 1 \end{pmatrix}$
(b) $\begin{pmatrix} 0 & 1 & 1 \\ -1 & 0 & 1 \\ -1 & -1 & 0 \end{pmatrix}$
(c) $\begin{pmatrix} 1 & 1 & 0 \\ 1 & 0 & -1 \\ 0 & -1 & -1 \end{pmatrix}$

Exercise 3.21

Use the `while` option to modify the solution to Exercise 3.6, so that the loop automatically terminates when $|x_k - N^{1/n}| < 10^{-8}$.

3.2.3 Case study: Goldbach's conjecture

In a letter to Leonhard Euler in 1742, the mathematician Christian Goldbach (1690–1764) made an elementary observation about prime numbers. Euler later re-expressed this observation in the form: every even number $N > 2$ is the sum of two primes. This very well known conjecture is called **Goldbach's conjecture**. Notably, the British publisher Tony Faber offered a \$1 000 000 prize to anyone who could prove the conjecture between 20 March 2000 and 20 March 2002, but the prize went unclaimed and the conjecture remains open.

We can use Maple to check this conjecture for some even numbers N, by splitting them into two primes, called its **prime summands**. One way of proceeding is as follows.

Suppose, for example, that we wish to split the number $N = 10$ into the sum of two primes. First construct an expression sequence `pxs` containing all the primes up to and including $N - 2$. To do this we use a loop with the `while` option.

```
>   N := 10:  pk := 2:  pxs := NULL:                    #1
>   for k while pk<=N-2 do                              #2
>     pxs := pxs,pk:                                    #3
>     pk := nextprime(pk):                              #4
>   end do:                                             #5
>   pxs;
```

$$2, 3, 5, 7$$

On line **#1** we define N, and initialize `pxs` and `pk`. On each iteration of the loop, `pk` is appended to `pxs`, after which `pk` is assigned to the next prime number. The loop terminates when $p_k > N - 2$. In this case, after the loop has executed, `pxs` has the value $2, 3, 5, 7$.

Next we need to take every element k in `pxs` and check whether $N - k$ is prime. If it is, then N is the sum of two primes, k and $N - k$.

```
>   for k in pxs do                                     #6
>     if isprime(N-k) then                              #7
>       print(k,N-k):                                   #8
>     end if:                                           #9
>   end do:                                             #10
```

$$3, 7$$
$$5, 5$$
$$7, 3$$

Note that there is some redundancy in this code, since $3, 7$ is output twice. One way to correct this is to demand that the prime summands $k, N - k$ are printed only if $k \geq N - k$.

Exercise 3.22

(a) Modify the preceding code so that the prime summands $k, N - k$ are printed only if $k \geq N - k$. Use your code to find the prime summands of 128.

(b) Modify this code so that instead of printing out the prime summands, it calculates the *number* of prime summands of N.

Exercise 3.23

The so-called **weak Goldbach conjecture** states that all odd numbers $N > 9$ can be written as the sum of three odd primes. In the following you are asked to construct code to find these summands, for the case $N = 25$.

(a) Modify lines **#1** to **#5** of the above code, so that $N = 25$ and `pxs` contains all odd primes less than or equal to $N - 6$.

(b) Use a pair of nested loops, each looping over the elements of `pxs`, to print out all prime triples which sum to N. Ensure that no triples $i, j, N - i - j$ are repeated, by demanding that they satisfy $i \geq j \geq N - i - j$.

The smallest odd prime is 3. Thus the largest possible prime that could be one of three primes summing to N would be $N - 6$.

3.3 Procedures

The last major programming construct covered in this unit is the **procedure**. Serious programming in any computer language involves writing procedures or their analogues, since their main purpose is to simplify programming, so that a complicated task can be built up from simpler sub-tasks.

In this section we explore various aspects of procedures, and apply them to a variety of problems to illustrate their use.

Basic ideas

A procedure is a self-contained Maple sub-program which can be executed many times, often within a larger program. It may have an input and an output, but otherwise is able to function autonomously. The input is an expression sequence, whose elements can be symbolic expressions, lists, matrices or any other Maple data types; likewise the output can be an expression sequence, a matrix, or even a Maple plot.

The simplest procedures simply mimic functions created using the `->` syntax. For example, the following procedure, called `dist1`, calculates the expression $\sqrt{x^2 + y^2}$, and is equivalent to the Maple function `(x,y)->sqrt(x^2+y^2)`.

```
>   dist1 := proc(x,y)
>     sqrt(x^2+y^2);
>   end proc;
```

As with loops, you must ensure that the procedure is in a single execution group.

$$dist1 := \mathbf{proc}(x, y)\, \mathrm{sqrt}(x^2 + y^2)\, \mathbf{end\ proc}$$

The output can be suppressed by changing the terminator after the `end proc` to the silent one (as for loops). Since there is generally no advantage to displaying this output, we will always terminate procedures with a silent terminator.

Output like this was encountered in *Unit 2* when using `dsolve` with the `numeric` option. This is because Maple actually returns a procedure when this command is input.

To execute this procedure with the values $x = -3$, $y = 4$ we input:

```
>   dist1(-3,4);
```

$$5$$

The following procedure converts Cartesian (x, y) to polar (r, θ) coordinates.

```
>   polar1 := proc(x,y)
>     sqrt(x^2+y^2),arctan(y,x):
>   end proc:
```

Note that the Maple function `arctan(y,x)` gives the polar angle of the point (x, y). Type `?arctan` for more information.

If we execute the procedure by typing

```
>   polar1(-3,4);
```

$$5,\ -\arctan\left(\frac{4}{3}\right) + \pi$$

we obtain an expression sequence as the output, which gives the polar coordinates r, θ.

Procedures are more general than functions, however, since the former can consist of any number of Maple statements, whereas the latter can only be a single expression or conditional statement. Let us construct a procedure consisting of several statements.

Suppose that we wish to calculate the angle θ between two vectors $\mathbf{v} = v_1\mathbf{i} + v_2\mathbf{j}$ and $\mathbf{w} = w_1\mathbf{i} + w_2\mathbf{j}$, as depicted in Figure 3.8. This can be obtained by using the formula

$$\cos\theta = \frac{\mathbf{v}\cdot\mathbf{w}}{|\mathbf{v}|\,|\mathbf{w}|} = \frac{v_1 w_1 + v_2 w_2}{\sqrt{v_1^2 + v_2^2}\sqrt{w_1^2 + w_2^2}}.$$

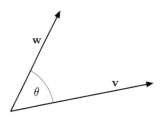

Figure 3.8

```
>   angle1 := proc(v1,v2,w1,w2)
>     vv := sqrt(v1^2+v2^2):
>     ww := sqrt(w1^2+w2^2):
>     vw := v1*w1+v2*w2:
>     arccos(vw/(vv*ww)):
>   end proc:
```

Here v1, v2, w1 and w2 are the components of **v** and **w**, respectively.

```
Warning, 'vv' is implicitly declared local to procedure 'angle1'
Warning, 'ww' is implicitly declared local to procedure 'angle1'
Warning, 'vw' is implicitly declared local to procedure 'angle1'
```

These warnings inform us that the values of the variables vv, ww and vw are local to the procedure angle1. This means that if these variables are assigned values inside the procedure, then these values cannot be accessed from outside, and vice versa. You should always suppress such warnings by explicitly declaring all variables used within a procedure, except for those in the argument list, as local (see line #2 below):

```
>   angle2 := proc(v1,v2,w1,w2)          #1
>     local vv,ww,vw:                     #2
>     vv := sqrt(v1^2+v2^2):              #3
>     ww := sqrt(w1^2+w2^2):              #4
>     vw := v1*w1+v2*w2:                  #5
>     arccos(vw/(vv*ww)):                 #6
>   end proc:                             #7
```

Note that the first terminator must come after the declaration of local variables on line #2.

So the angle between the vectors $\mathbf{j}$ and $\mathbf{i}+\mathbf{j}$ is:

```
>   angle2(0,1,1,1);
```

$$\frac{\pi}{4}$$

If we attempt to access the value of the variable vv from outside angle2 by typing

```
>   vv;
```

$$vv$$

we find that it is unassigned, because as mentioned, it is local to angle2.

The local nature of variables within procedures is actually very useful, since a variable with a simple name such as vv, or x, is natural to use in different contexts. If (say) x is used as a local variable in a procedure, then x can be used elsewhere in the code (perhaps in a different procedure), and Maple will treat it as a *different variable*; it will not confuse them.

A global declaration for procedure variables also exists, and you will see examples of its use in *Units 3* and *4* of Block B.

131

It is important to note that although there are four lines (#3 to #6) in the body of `angle2`, Maple only returns the output from line #6. This is because, by default, the output returned by a procedure is always that from the last statement to be executed. (We will see how to modify this behaviour using **return** shortly.)

As mentioned at the beginning of this section, the main purpose of procedures is to split complicated tasks into sub-tasks. By way of illustration, suppose that we want to calculate all the angles of the triangle with vertices $(0,0)$, (v_1, v_2) and (w_1, w_2). The easiest way to do this is to use the procedure `angle2` to calculate the angles θ_1 between $\mathbf{v}$ and $\mathbf{w}$, and θ_2 between $-\mathbf{v}$ and $\mathbf{w} - \mathbf{v}$ (see Figure 3.9), as in the following procedure:

The third angle is obtained by calculating $\pi - \theta_1 - \theta_2$.

```
>   angles := proc(v1,v2,w1,w2)          #1
>     local theta1,theta2:               #2
>     theta1 := angle2(v1,v2,w1,w2):      #3
>     theta2 := angle2(-v1,-v2,w1-v1,w2-v2):  #4
>     theta1,theta2,Pi-theta1-theta2:    #5
>   end proc:                            #6
>   angles(0,1,1,1);
```

$$\frac{\pi}{4}, \frac{\pi}{2}, \frac{\pi}{4}$$

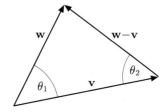

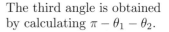

Figure 3.9

(The vectors here are $(0,1)$ and $(1,1)$; they are not the general vectors shown in Figure 3.9.) The output on line #5 is the expression sequence $\theta_1, \theta_2, \pi - \theta_1 - \theta_2$. Note that as `angles` depends on `angle2`, it is necessary to execute `angle2` before `angles`, and it is advisable to combine both procedures into a single execution group.

Example 3.6

Construct a procedure `f(N)`, with input N and output

$$f(N) = \prod_{k=2}^{N} \frac{1}{1 + \sin k}.$$

Solution

Note that in solving Exercise 3.3(c), you actually calculated $f(10)$ by means of a loop. The first thing to do is to modify this loop so that the upper limit becomes the variable N:

```
>   N := 10:                    #1
>   pr := 1:                    #2
>   for k from 2 to N do        #3
>     pr := pr*1/(1+sin(k)):    #4
>   end do:                     #5
>   evalf(pr);                  #6
```

$$30.04124664$$

Next, this code is turned into a procedure by replacing line #1 with the two lines `F := proc(N)` and `local k,pr:`, and putting **end proc:** after line #6. So the solution is

```
>  F := proc(N)
>    local k,pr:
>    pr := 1:
>    for k from 2 to N do
>      pr := pr*1/(1+sin(k)):
>    end do:
>    evalf(pr):
>  end proc:
```
♦

Exercise 3.24

Create procedures to mimic the following functions.

(a) $r(x, y, z) = \sqrt{x^2 + y^2 + z^2}$

(b) $S(N) = \sum_{k=1}^{N} \frac{(-1)^k}{k}$

(Check your result against that for Exercise 3.3(a).)

(c) $\text{fact}(n, k) = n \times (n - k) \times (n - 2k) \times \cdots \times R,$

where n and k are positive numbers, and R is the last number in the product to be greater than or equal to 1.

(Check your result against that for Exercise 3.5.)

Exercise 3.25

Create a procedure to calculate

$$C2(N) = \prod_{k=2}^{N} \left(1 - \frac{1}{(p_k - 1)^2}\right),$$

where p_k is the kth prime number.

As $N \to \infty$, $C2(N)$ tends to a constant C_2, which is related to the Goldbach conjecture (as you will see in Exercise 3.30). Plot a graph of C_2 for $50 \leq N \leq 500$, and thereby show that $C_2 \simeq 0.6602$.

Procedure arguments

We have commented on the fact that all variables used within a procedure, except for those in the argument list, should be declared as `local`. The question arises as to what status the variables in the argument list have. This is easily demonstrated by considering the following problem.

Suppose that we wish to construct a procedure, with input N and x_0, and with output x_N, defined by Equation (3.1) on page 112. Following the solution to Exercise 3.1, we might be tempted to write:

```
>  xN1 := proc(x,N)
>    local k:
>    for k to N do  x := x/2+5/(2*x):  end do:          #1
>  end proc:
```

Unfortunately, if we now try to find (say) x_4 by inputting `xN1(1,4);`, we obtain an error message:

```
>  xN1(1,4);
Error, (in xN1) illegal use of a formal parameter
```

This is because Maple evaluates variables which occur in the argument list. So when we input `xN1(1,4);`, Maple interprets line #1 as

```
    for k to 4 do  1 := 1/2+5/(2*1):  end do:
```

To put it another way, the command `xN1(1,4);` is telling Maple to treat x and N as 1 and 4, respectively, *throughout the execution of the procedure*, and it cannot do this if the procedure tries to re-assign a value for x.

This can be easily corrected by assigning variables in the argument list to local variables, then using these:

```
>  xN2 := proc(x0,N)
>    local k,x:
>    x := x0:
>    for k to N do  x := x/2+5/(2*x):  end do:
>  end proc:
>  xN2(1,4);
```

$$\frac{2207}{987}$$

Here x0 is assigned to the local variable x, and it is this variable which is used in all further assignments.

The key point to remember here is: *a variable which occurs in the argument list should never be used on the left-hand side of an assignment statement.*

Exercise 3.26

Write a procedure `iter` with input N, n, M and x_0, and with output x_M, where

$$x_{k+1} = \frac{1}{n}\left(\frac{N}{x_k^{n-1}} + (n-1)x_k\right) \quad \text{for } k = 0, 1, 2, \ldots.$$

Check the output from your procedure against the results from Exercise 3.6.

The return *statement*

As mentioned, by default, the output returned by a procedure is that of the last statement to be executed within it. However, it is sometimes necessary to terminate the procedure prematurely and return a different output.

For example, the procedure `polar1` (on page 130), to convert Cartesian to polar coordinates, is not entirely satisfactory because it does not work correctly for $x = y = 0$, where the polar angle θ is not defined (`polar1(0,0)` returns the value $0,0$). We therefore modify it as follows:

```
> polar2 := proc(x,y)                                    #1
>   if x=0 and y=0 then                                  #2
>     print("polar angle not defined"):                  #3
>     return:                                             #4
>   end if:                                               #5
>   sqrt(x^2+y^2),arctan(y,x):                            #6
> end proc:                                               #7
```

Now we obtain

```
> polar2(0,0);
```
$$\text{"polar angle not defined"}$$

If both the arguments (x and y) are zero, then the conditional on line #2 is true, so Maple prints "polar angle not defined" and then executes the **return** statement, which causes Maple to terminate the execution of the procedure.

More generally, if the **return** statement is used in the form

    return *expression sequence*

then it returns the *expression sequence* and then terminates the execution of the procedure. For example, the following procedure identifies the first position of the element el in the list L.

```
> position := proc(el,L)                                 #1
>   local k:                                             #2
>   for k to nops(L) do                                  #3
>     if L[k]=el then return k:  end if:                 #4
>   end do:                                              #5
>   print(el||" not an element of the list")            #6
> end proc:                                              #7
> position(aa,[acc,ee,aa,ff,g]);
```
$$3$$

```
> position(abc,[acc,ee,aa,ff,g]);
```
$$abc \text{ not an element of the list}$$

On line #4, we loop over the elements of L, checking if each is equal to el. If one is, then the **return** statement returns its position and terminates the procedure. If no element of L is equal to el, then the procedure prints an appropriate message.

Exercise 3.27

Create a procedure `polar3`, whose input is the Cartesian coordinates x, y, z and whose output is an expression sequence r, θ, ϕ, given by the spherical polar coordinates (see Figure 3.10)

$$r = \sqrt{x^2 + y^2 + z^2}, \quad \theta = \arctan(\sqrt{x^2 + y^2}, z), \quad \phi = \arctan(y, x).$$

Ensure that your procedure outputs appropriate error messages if $x = y = 0$, $z \neq 0$ (ϕ not defined), or $x = y = z = 0$ (θ and ϕ not defined).

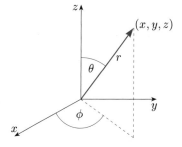

Figure 3.10

Declaring data types

When writing lengthy Maple programs, it is often necessary to call several procedures and to work with many different **data types**. Calling procedures with the wrong data type in the argument list can lead to unpredictable behaviour. For example, the procedure `null1` (below) takes a 2×2 matrix **M** and sets its diagonal elements to zero.

```
>  null1 := proc(M)
>     local k:
>     for k from 1 to 2 do  M[k,k] := 0:  end do:
>     M:
>  end proc:
>  H := Matrix([[1,2],[3,4]]);
```

$$H := \begin{bmatrix} 1 & 2 \\ 3 & 4 \end{bmatrix}$$

```
>  null1(H);
```

$$\begin{bmatrix} 0 & 2 \\ 3 & 0 \end{bmatrix}$$

However, if we (accidentally) call `null1` with just a name W in its argument list, then nothing much happens:

```
>  null1(W);
```

$$W$$

Maple allows us to define the intended data types of a procedure's parameters as follows:

```
>  null2 := proc(M::Matrix(2,2))
>     local k:
>     for k from 1 to 2 do  M[k,k] := 0:  end do:
>     M:
>  end proc:
```

`::` was used in *Unit 1* with the `assume` command.

Maple will now check that we are calling this procedure with the correct data type. It behaves identically to `null1` on 2×2 matrices, but when we try to call it with a different data type, we get an explicit error message:

```
>  null2(W);
```

`Error, invalid input: null2 expects its 1st argument, M, to be of type Matrix(2,2), but received W`

The data types that can be declared in the argument list are the same as those used in other applications of Maple, together with a couple of additions (see the appropriate Maple Help page for more details, by inputting `?type`). Some common data types are displayed in Table 3.1.

Table 3.1   Some data types

Type name	Data type
Matrix	Matrix
Vector	Vector
list	list
set	set
integer	integer
positive	positive
negative	negative
even	even
odd	odd
prime	prime
complex	complex
imaginary	imaginary
realcons	real and constant
posint	positive integer
negint	negative integer
range	range, e.g. a..b
mathfunc	function, e.g. cos

Exercise 3.28

Create a procedure `angle3`, with the vectors $\mathbf{v} = v_1\mathbf{i} + v_2\mathbf{j} + v_3\mathbf{k}$ and $\mathbf{w} = w_1\mathbf{i} + w_2\mathbf{j} + w_3\mathbf{k}$ as input and with the angle θ between them as output.

[*Hint*: Recall that

$$\cos\theta = \frac{\mathbf{v}\cdot\mathbf{w}}{|\mathbf{v}|\,|\mathbf{w}|} = \frac{v_1 w_1 + v_2 w_2 + v_3 w_3}{\sqrt{v_1^2 + v_2^2 + v_3^2}\sqrt{w_1^2 + w_2^2 + w_3^2}}.]$$

Your procedure should issue an appropriate error message if $\mathbf{v}$ and $\mathbf{w}$ are not three-dimensional vectors.

Outputting plots

Procedures can be made to output plots. As an example, consider the solution of the differential equation

$$\frac{dy(t)}{dt} + y(t)^2 = t,$$

subject to the initial condition $y(0) = a$. In Subsection 2.6.3 of *Unit 2*, this differential equation was solved numerically for $a = 1$, using `dsolve`, and a graph of the solution was plotted. You should remind yourself of how the solution and plot were obtained before proceeding further.

Here let us define a procedure `gr(a::realcons,rnge::range,optns)` which automatically produces a plot for initial condition $y(0) = a$, over the range given by the parameter `rnge`, using plot options given by the parameter `optns`.

`::range` checks that `rnge` is of data type range.

```
>   gr := proc(a::realcons,rnge::range,optns)
>       local eq,ic,sol,fy,t0:
>       eq := D(y)(t)+y(t)^2=t;
>       ic := y(0)=a:
>       sol := dsolve({eq,ic},numeric):
>       fy := t0->eval(y(t),sol(t0));
>       plot(fy,rnge,optns);
>   end proc:
```

Now if we input

```
>   gr(1,0..3,colour=red);
```

we obtain the same plot as in Figure 2.18 of *Unit 2*.

However, if we want to compare the solution for three different values of the initial condition, say $a = 0, 1, 2$, by displaying all three graphs on the same plot, we input:

```
>   plots[display]([gr(0,0..3,colour=red),
       gr(1,0..3,colour=black),gr(2,0..3,colour=grey)]);
```

The output is shown in Figure 3.11.

Here all graphs are plotted in the range $0 \le t \le 3$. The graph corresponding to $a = 0$ is the red line, that for $a = 1$ is the black line, and that for $a = 2$ is the grey line.

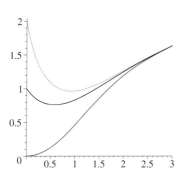

Figure 3.11

137

Exercise 3.29

(a) Create a procedure `gr1(a,rnge,optns)` which automatically produces a plot of the solution of the differential equation

$$\frac{dy(x)}{dx} = \cos(\pi x \, y(x))$$

for initial condition $y(0) = a$. The range of the plot should be given by the parameter `rnge`, and any plot options given by the parameter `optns`. You should declare the data types for `a` and `rnge`.

[*Hint*: Solve this differential equation numerically.]

(b) On the same graph, plot three solutions for the initial conditions $a = 1, 2, 3$, over the range $0 \leq x \leq 6$.

Exercise 3.30

This exercise is about the number of ways that an even number can be expressed as the sum of two primes. It builds on the code that you produced in Subsection 3.2.3 on Goldbach's conjecture.

(a) Use the code you produced in Exercise 3.22(b) to create a procedure `numprime(N)`, whose input is an even number $N > 2$, and whose output is $\nu(N)$, the number of prime summands of N.

Ensure that your procedure outputs an appropriate error message if N is not even or if $N \leq 2$.

(b) Use your procedure to plot $\nu(N)$ for $4 \leq N \leq 2000$ (use the plot option `style=point`).

(c) On the same graph as in part (b), plot the function $C_2 \dfrac{N}{(\ln(N))^2}$, where $C_2 \simeq 0.6602$ is the constant calculated in Exercise 3.25.

You should notice that $C_2 \dfrac{N}{(\ln(N))^2}$ provides an accurate estimate of the smallest number of prime summands of N.

3.4   End of unit exercises

These exercises are all quite substantial, and you may wish to leave some of them for later revision. Exercise 3.37 is particularly challenging.

Exercise 3.31

Marin Mersenne (1588–1648) conjectured that the only primes p for which $M_p = 2^p - 1$ is prime are $2, 3, 5, 7, 13, 17, 19, 31, 67, 127, 257$. Was he correct? Can you find any Mersenne primes larger than M_{257}?

Exercise 3.32

Use first a conditional and then `piecewise` to create the double top-hat function

$$T(x) = \begin{cases} 1 & \text{if } x < 1, \\ 2 & \text{if } 1 \le x \le 2, \\ 1 & \text{if } 2 < x < 3, \\ 2 & \text{if } 3 \le x \le 4, \\ 1 & \text{if } x > 4, \end{cases}$$

and plot it for $-2 \le x \le 6$.

Exercise 3.33

The nth spherical Legendre polynomial, $P_n(x)$, is defined recursively by

$$(n+1)P_{n+1} = (2n+1)xP_n - nP_{n-1} \quad \text{for } n = 1, 2, 3, \ldots,$$

where $P_0(x) = 1$ and $P_1(x) = x$.

(a) Use a loop construction to determine $P_0(x), \ldots, P_5(x)$.

(b) Show that, for $n = 1, 2, \ldots, 5$, these polynomials satisfy **Legendre's equation**

$$(1 - x^2)\frac{d^2}{dx^2}P_n - 2x\frac{d}{dx}P_n + n(n+1)P_n = 0.$$

This differential equation commonly occurs in applied mathematics when problems are expressed in spherical polar coordinates. It is named after the French mathematician Adrien-Marie Legendre (1752–1833).

(c) For $n = 1, 2, \ldots, 5$, verify that

$$P_n(x) = \frac{1}{2^n\, n!}\frac{d^n}{dx^n}(x^2 - 1)^n.$$

(d) Verify that $P_0(\cos\theta) = 1$, $P_1(\cos\theta) = \cos\theta$, $P_2(\cos\theta) = \frac{3}{4}\cos 2\theta + \frac{1}{4}$, $P_3(\cos\theta) = \frac{5}{8}\cos 3\theta + \frac{3}{8}\cos\theta$, and find similar expressions for P_4 and P_5 in terms of $\cos n\theta$ (n integer). [Hint: Use `combine`.]

(e) Show that, for $i, j = 0, 1, \ldots, 5$,

$$\int_{-1}^{1} P_i(x)\, P_j(x)\, dx = \begin{cases} 0 & \text{if } i \ne j, \\ \frac{2}{2i+1} & \text{if } i = j. \end{cases}$$

Exercise 3.34

If three integers $l, m, N \ge 1$ satisfy $N = l^2 + m^2$, we say that N is the sum of two squares.

(a) Use a pair of nested loops and a conditional to determine if an integer N can be written as the sum of two squares. Test your code for $N = 10$.

(b) Convert this code into a procedure `psum(N)`, whose input is a positive integer N and whose output is the NULL sequence if N is not the sum of two squares, and otherwise is the sequence $[n_1, m_1], [n_2, m_2], \ldots, [n_M, m_M]$ of the pairs of numbers whose squares add to N. Your procedure should issue an appropriate error message if N is not a positive integer.

(c) Use `psum(N)` to find the first integer N with four pairs of numbers whose squares sum to it.

Exercise 3.35

If a function $f(x)$ is defined over a range $a \leq x < b$, then its **periodic extension** $F(x)$ is obtained by translating copies of itself along the x-axis by multiples of $P = b - a$. The result is a periodic function with period P, such that $F(x) = f(x)$ for $a \leq x < b$. For example, in Figure 3.12, we define the function $f(x) = x$ over the range $-1 \leq x < 2$ (thick line), and the periodic extension $F(x)$ (thin line) is obtained by translating copies of this line segment along the x-axis by $x = 0, \pm 3, \pm 6, \ldots$.

Formally, the periodic extension of a function $f(x)$ defined over a range $a \leq x < b$ is given by

$$F(x) = f(x - nP)$$

where $P = b - a$ and n is the number of multiples of $b - a$ by which x differs from a value in the range $[a, b)$. More neatly:

$$F(x) = f\left(x - P\left[\frac{x-a}{P}\right]\right) \quad \text{where } P = b - a.$$

(a) Create a procedure per(x,f,a,b), which takes a function f defined over the range $[a, b)$ and returns the value $F(x)$ for any x.

(b) For $f(x) = x$, $a = -1$, $b = 2$, use per to plot the periodic extension $F(x)$ for $-5 \leq x < 5$.

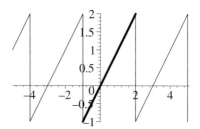

Figure 3.12   The function $f(x)$ (thick line) and its periodic extension $F(x)$ (thin line)

Remember that $[\frac{x-a}{P}]$ means the integer part of $\frac{x-a}{P}$.

Exercise 3.36

In this exercise we ask you to apply conditionals to the problem of finding repeated sub-sequences in the digits of π.

The set of one-digit sub-sequences is simply the ten digits 0–9. The first repeated digit (1) occurs when π is expressed to three decimal places: 3.1**41**.

The set of two-digit sub-sequences is the one hundred digit pairs $00, 01, 02, \ldots, 98, 99$. The 22nd two-digit sub-sequence (26) is the first to repeat: $\pi = 3.141592$**6**535897932384**626**$\ldots$.

In Exercise 3.10 you constructed a loop to generate the three-digit sub-sequences in the expansion of π i.e. $314, 141, 415, 159, \ldots$.

(a) Alter this code so that each three-digit sub-sequence becomes an element of an expression sequence Xs := 314,141,415,159,....

(b) The simplest way to check if the elements of an expression sequence are repeated is to convert it to both a list and a set and to compare the number of elements in each.

For example, given the expression sequence Xs2 := 123,234,345,456,123, the list Ls2:=[Xs2] has the value [123,234,345,456,123], while the set Set2:={Xs2} has the value {123,234,345,456}, because repeated elements are not conserved for sets. Hence nops(Ls2) = 5 and nops(Set2) = 4.

Use this idea to verify that the 62nd 3-digit sub-sequence (592) is the first to repeat.

(c) Modify your code so that it finds the first repeated P-digit sub-sequence (for any positive integer P). Test your code on 1-, 2- and 3-digit sub-sequences. Which is the first repeated 5-digit sub-sequence?

Exercise 3.37

In this question we ask you to calculate the volume of an N-dimensional ball.

(a) The volumes V_2 and V_3 of a 2-dimensional and a 3-dimensional ball are given by the integrals

A 2-dimensional ball is a disk.

$$V_2 = \int_{-r}^{r} \left(\int_{-\sqrt{r^2 - x_2^2}}^{\sqrt{r^2 - x_2^2}} 1 \, dx_1 \right) dx_2,$$

$$V_3 = \int_{-r}^{r} \left(\int_{-\sqrt{r^2 - x_3^2}}^{\sqrt{r^2 - x_3^2}} \left(\int_{-\sqrt{r^2 - x_2^2 - x_3^2}}^{\sqrt{r^2 - x_2^2 - x_3^2}} 1 \, dx_1 \right) dx_2 \right) dx_3.$$

Use Maple to verify that $V_2 = \pi r^2$ and $V_3 = \frac{4}{3} \pi r^3$.

[*Hint*: You must inform Maple that $r > 0$.]

(b) By generalizing the integrals for V_2 and V_3, write code to calculate V_N for any integer N.

Summary of Unit 3

The main commands and constructs introduced in this unit are as follows.

`for` loop	for looping over Maple statements
`next`	for advancing to the next loop iteration
`break`	for terminating a loop prematurely
`NULL`	the expression sequence with no elements
`if`	for writing conditionals
`else, elif`	for writing conditionals with clauses
`and, or, not`	logical operators
`while`	for terminating a loop conditionally
`proc`	for writing a procedure
`local`	for declaring local variables
`return`	for terminating a procedure prematurely
`::`	for data type declaration
`piecewise`	for defining piecewise functions
`print`	for producing output from silent loops, and from procedures

Learning outcomes

After studying this unit you should be able to:

- understand and write basic loops;
- use loops to calculate expression sequences and expressions defined iteratively;
- advance to the next loop iteration using `next`;
- terminate loops prematurely using `break` or `while`;
- understand and write conditional statements, using the clauses `else` and `elif`;
- use conditionals and `piecewise` to construct piecewise-defined functions;
- loop over the elements of a list;
- use nested loops, for example to construct matrices;
- construct basic procedures;
- understand the nature of local variables;
- use the `return` statement to terminate the execution of a procedure;
- know how to declare data types.

Solutions to Exercises

Solution 3.1

(a) On the first iteration of the loop $m = 3$, while on the second iteration $m = 6$. There is no third iteration as the next value of the index would be $m = 9$. So we expect the output:

```
>   for m from 3 to 7 by 3 do
>     m^2;
>   end do;
```
$$9$$
$$36$$

(b) On the first iteration $m = 3$. There is no second iteration as the next value of the index would be $m = 8$. So we expect the output:

```
>   for m from 3 to 7 by 5 do
>     m^2;
>   end do;
```
$$9$$

Solution 3.2

(a) This loop simply outputs the numbers $0, 1, 2, 3, 4$.

(b) In this loop, the index takes the values $k = 2, 4$, so the input and output are:

```
>   for k from 2 to 5 by 2 do sin(k):  end do;
```
$$\sin(2)$$
$$\sin(4)$$

(c) In this loop, the index takes the values $k = 10, 7, 4$, so the input and output are:

```
>   for k from 10 to 2 by -3 do
>     k^2:
>     print(k,k^2):
>   end do:
```
$$10, 100$$
$$7, 49$$
$$4, 16$$

Note that the loop ends with a silent terminator, so there is output only from the **print** statement.

After executing these loops, the index k has the values: (a) $k = 6$, (b) $k = 6$, (c) $k = 1$.

Generally, when a loop

```
   for k from n1 to n2 by n3 do
```

is executed, the index k takes the values $n_1, n_1 + n_3, n_1 + 2n_3, \ldots$ until its value exceeds n_2. It retains this final value after the loop has executed.

Solution 3.3

(a) The following loop constructs the sum:

```
>   sm := 0:                              #1
>   for k from 1 to 100 by 1 do           #2
>     sm := sm+(-1)^k/k:                   #3
>   end do:                               #4
```

On line **#1** we initialize the sum **sm** to zero. Then on every iteration ($k = 1, 2, \ldots, 100$) of the loop we add a term $(-1)^k/k$ to **sm**.

The value of the sum is given by

```
>   evalf(sm);
```
$$-0.6881721793$$

You can check the result using **add**:

```
>   evalf(add((-1)^k/k,k=1..100));
```
$$-0.6881721793$$

(b) Here we can proceed as follows:

```
>   sm := 0:
>   for k from 5 to 20 by 1 do
>     sm := sm+1/(1-k^2):
>   end do:
>   evalf(sm);
```
$$-0.1761904762$$

You can check the result using **add**:

```
>   evalf(add(1/(1-k^2),k=5..20));
```
$$-0.1761904762$$

(c) In this case:

```
>   pr := 1:                              #1
>   for k from 2 to 10 by 1 do            #2
>     pr := pr*1/(1+sin(k)):              #3
>   end do:                               #4
```

On line **#1** we initialize the product **pr** to 1. Then on every iteration ($k = 2, 3, \ldots, 10$) of the loop it is multiplied by $1/(1 + \sin k)$.

The final value of **pr** is

```
>   evalf(pr);
```
$$30.04124664$$

You can check the result using **mul**:

```
>   evalf(mul(1/(1+sin(k)),k=2..10));
```
$$30.04124664$$

Solution 3.4

(a) `for k from -3 to 7 by 1 do`

(b) `for k from 1 to 7 by 3 do`

(c) `for k from -1 to infinity by 2 do`

Solution 3.5

(a) Set up the loop:

```
> pr := 1:
> for k to 10 do
>    pr := pr*k:
> end do:
> pr;
                3628800
```

We can check the result:

```
> 10!;
                3628800
```

(b) This loop is set up as follows:

```
> pr := 1:
> for k from 2 to 20 by 3 do        #1
>    pr := pr*k:
> end do:
> pr;
                4188800
```

Alternatively, for line #1 you could have chosen

```
> for k from 20 to 2 by -3 do
```

Solution 3.6

In each case we perform eight iterations of the formula, and compare x_k with $N^{1/n}$. (If that turns out not to be enough, we can always increase the number of iterations.)

(a) For this part:

```
> x := 1:  N := 3:  n := 2:  #initial values
> for k to 8 do
>    x := (N/x^(n-1)+(n-1)*x)/n:   #formula
>    print(k,evalf(x-N^(1/n))):    #output
> end do:
            1, 0.267949192
            2, 0.017949192
            3, 0.000092049
            4, 0.2 10^{-8}
            5, 0.
            6, 0.
            7, 0.
            8, 0.
```

So four iterations are required.

(b) Replace the first line in the code with

```
> x := 1:  N := 7:  n := 4:
            1, 0.873423438
            2, 0.360423438
            3, 0.086745123
            4, 0.006368285
            5, 0.000037156
            6, 0.1 10^{-8}
            7, 0.
            8, 0.
```

So six iterations are required.

(c) Replace the first line in the code with

```
> x := 1:  N := 23:  n := 3:
            1, 5.489466353
            2, 2.822088576
            3, 1.172250983
            4, 0.308873282
            5, 0.029272052
            6, 0.000297216
            7, 0.31 10^{-7}
            8, 0.
```

So eight iterations are required.

Solution 3.7

(a) If we use the loop

```
> restart:  f := 1:
> for k from 2 to 8 do
>    f := 1/(1-x*f):
> end do:
```

to calculate f_8, then we obtain the result

```
> f;
```

$$\cfrac{1}{1-\cfrac{x}{1-\cfrac{x}{1-\cfrac{x}{1-\cfrac{x}{1-\cfrac{x}{1-\cfrac{x}{1-x}}}}}}}$$

We can use `simplify` to put f into rational form:

```
> f := simplify(f);
```

$$f := -\frac{-1+6x-10x^2+4x^3}{1-7x+15x^2-10x^3+x^4}$$

(b) Let us represent f_∞ by F:

```
> eq := F=1/(1-x*F);
```

$$eq := F = \frac{1}{1-xF}$$

```
> Fsol := solve(eq,F);
```

$$Fsol := \frac{1+\sqrt{1-4x}}{2x}, \; \frac{-1+\sqrt{1-4x}}{2x}$$

So there are two possible solutions for f_∞. We can plot these together with f_8 over the suggested ranges for x and y:

```
> plot([f,Fsol],x=-20..0,-1..1,
    colour=[black,grey,red]);
```

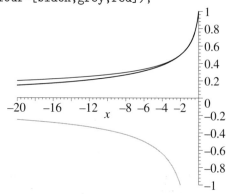

Observe that f_8 (black line) is a good approximation to one of these solutions (`Fsol[2]`) for $x < 0$.

Solution 3.8

A simple method is to express π to 10 006 decimal places, multiply by 10^{9999}, and read off the first five digits after the decimal point:

```
>  N := 10005:  Digits := N+1:
   pi1 := evalf(Pi):
>  frac(pi1*10^(N-6));
```
$$0.856672$$

So the 10 001st to 10 005th digits are $8, 5, 6, 6, 7$.

Note that this code is general, and will generate the $(N-4)$th to Nth digits for any $N > 4$.

Solution 3.9

Generating the first digit is easy:

```
>  restart:  N := 5:  pi1 := evalf[N+1](Pi);
```
$$\pi 1 := 3.14159$$

```
>  dig := floor(pi1);
```
$$dig := 3$$

We can generate the next digit by subtracting the integer part of π from π, and multiplying by 10:

```
>  pi1 := 10*(pi1-dig);
```
$$\pi 1 := 1.41590$$

```
>  dig := floor(pi1);  pi1 := 10*(pi1-dig);
```
$$dig := 1$$
$$\pi 1 := 4.15900$$

Now repeat the procedure, using a loop to automate it:

```
>  restart:  N := 5:  pi1 := evalf[N+1](Pi):
>  for k from 1 to 5 do
>    dig := floor(pi1):
>    pi1 := 10*(pi1-dig):
>  end do;
```

The same output as for Code 3.1 on page 117 is obtained.

Solution 3.10

One method is to modify Code 3.1 on page 117 in the following way:

```
>  restart:
   N := 5:  pi1 := evalf[N+1](Pi);        #1
>  for k from 1 to N-2 do                 #2
>    dig := floor(100*pi1):               #3
>    pi1 := evalf[N+1](10*frac(pi1)):     #4
>  end do;
```
$$\pi 1 := 3.14159$$
$$dig := 314$$
$$\pi 1 := 1.41590$$
$$dig := 141$$
$$\pi 1 := 4.15900$$
$$dig := 415$$
$$\pi 1 := 1.59000$$

Here, and in a few other places, we have moved all the output together for easier comparison. The main modification, on line #3, is to multiply pi1 by 100.

This means that the integer part of 100*pi1 always contains 3 digits. The only other modification is on the upper bound (in line #2), which stops the loop from extracting too many digits from pi1.

Solution 3.11

Using Code 3.2 on page 118, with N:=1000, we obtain the following value for num:

```
>  num;
```
$$[93, 116, 103, 103, 93, 97, 94, 95, 101, 105]$$

which shows that there are 105 occurrences of 9 in the first 1000 digits of π.

Solution 3.12

Recall that the exponential number e is denoted by exp(1) in Maple.

We first calculate the frequency of each digit of e when expressed to 1999 decimal places. To do this we simply change line #1 of Code 3.2 on page 118 to

```
>  restart:
   N := 1999:  e1 := evalf[N+1](exp(1)):
```

The remainder of the code remains essentially the same (substitute e1 for pi1), and when it is executed num has the value

$$[196, 190, 208, 201, 201, 197, 204, 198, 202, 202]$$

So 1 occurs 190 times.

Next we calculate the frequency of the digits of e when expressed to 3000 decimal places, by changing line #1 to

```
>  restart:
   N := 3000:  e1 := evalf[N+1](exp(1)):
```

This time num has the value

$$[301, 280, 301, 304, 291, 296, 315, 307, 295, 310]$$

So 1 occurs 280 times.

Thus there are $280 - 190 = 90$ occurrences of 1 between the 2000th and 3000th digits, inclusive, of the exponential number e.

Solution 3.13

(a) Let N denote the number of hexadecimal digits required. Then, following the hint in the question, we calculate π to 7 significant figures:

```
>  restart:  N := 5:
>  M := floor(log[10](16^N));
   pi1 := evalf[M+1](Pi);
```
$$M := 6$$
$$\pi 1 := 3.141593$$

We now extract the hexadecimal digits of π in an analogous manner to the way that we extracted the decimal digits (cf. Code 3.1):

```
>  for k from 1 to N do
>    dig := floor(pi1):
>    pi1 := evalf[M+1](16*frac(pi1)):
>  end do;
```

$$dig := 3$$
$$\pi1 := 2.265488$$
$$dig := 2$$
$$\pi1 := 4.247808$$
$$dig := 4$$
$$\pi1 := 3.964928$$
$$dig := 3$$
$$\pi1 := 15.43885$$
$$dig := 15$$
$$\pi1 := 7.02160$$

(Note that this code is Code 3.1 with `evalf[N+1]` replaced by `evalf[M+1]` and all occurrences of 10 replaced by 16.)

So $c_0 = 3$, $c_1 = 2$, $c_2 = 4$, $c_3 = 3$, $c_4 = 15$.

It is easy to check that this is correct (in this case to 5 significant figures):

```
>   evalf(3*16^0+2/16+4/16^2+3/16^3+15/16^4);
                    3.141586304
```

(b) We modify Code 3.2 in a similarly straightforward manner:

```
>   restart:  N := 10000:
>   M := floor(log[10](16^N)):
    pi1 := evalf[M+1](Pi):
>   num := [0$16]:
>   for k from 1 to N do
>     dig := floor(pi1):
>     pi1 := evalf[M+1](16*frac(pi1)):
>     num[dig+1] := num[dig+1]+1:
>   end do:
>   dat := [seq([k-1,num[k]/10000],k=1..16)]:
>   plot(dat,view=[0..15,0..0.1]);
```

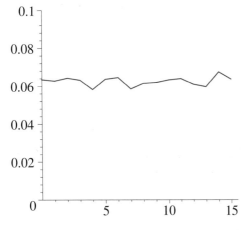

Once again, note that this code is Code 3.2 with all occurrences of 10 replaced by 16.

We observe that each hexadecimal digit occurs roughly 1/16th of the time.

It is an unproven conjecture that, no matter what base π is expressed in, the digits always occur with equal frequency.

Solution 3.14

The following loop checks each integer $1 \le k \le 50$ to see if it is prime or divisible by 2 or by 3. If it is, then it moves to the next integer, without doing anything, via the **next** statement. If it isn't, then the integer is appended to the sequence Xs.

```
>   Xs := NULL:
>   for k to 50 do
>       if isprime(k) or frac(k/2)=0
           or frac(k/3)=0 then
>          next
>       end if:
>       Xs := Xs,k:
>   end do:
>   Xs;
```
$$1, 25, 35, 49$$

Solution 3.15

The following code loops over each prime number $p_k \ge 2$. If $p_k \le 5000$ it moves on to the next prime, and if $p_k \ge 5050$ the loop terminates. Otherwise, p_k is appended to the sequence Xs.

```
>   Xs := NULL:
>   for k do
>       pk := ithprime(k):
>       if pk<=5000 then next   end if:
>       if pk>=5050 then break   end if:
>       Xs := Xs,pk:
>   end do:
>   Xs;
```
$$5003, 5009, 5011, 5021, 5023, 5039$$

There are other ways of achieving this result; one alternative is discussed on page 126.

Solution 3.16

(a) Define the function $f(n) = n^2 - n + 41$:

```
>   f := n->n^2-n+41:
```

Now loop over every integer $n \ge 1$, and if $f(n)$ is not prime then print n and exit the loop.

```
>   for n do
>       if not isprime(f(n)) then
>          print(n):
>          break:
>       end if:
>   end do:
```
$$41$$

So the first non-prime value of $f(n)$ occurs when $n = 41$. (It is easy to see that $f(41) = 41^2$.)

(b) Use exactly the same loop as in part (a), but for the function $f(n) = n^2 - 79n + 1601$. The result is that $f(n)$ is prime for $1 \le n \le 79$. (For $n = 80$ you may wish to check that $f(80) = 41^2$.)

Solution 3.17

```
>  f := (x,y)->
>     if x>=y then [x,y]
>     else [y,x]
>     end if:
```

Check this works:

```
>  f(1,7);
```

$$[7, 1]$$

Note that you could alternatively use `sort` to do this (type `?sort` for more details).

Solution 3.18

```
>  restart:
>  f := (a,b,c)->
>     if b^2-4*a*c=0 then
>          print("one real solution"):
>     elif b^2-4*a*c>0 then
>          print("two real solutions"):
>     elif b=0 then
>          print("two imaginary solutions"):
>     else print("two complex solutions
>               (not pure imaginary)"):
>     end if:
```

For example, for $3x^2 + 2x + 2 = 0$, we have

```
>  f(3,2,2);
```

"two complex solutions (not pure imaginary)"

Solution 3.19

(a) Using conditionals, f is defined as follows:

```
>  f := x->
>     if evalf(abs(sin(x)))<=0.5 then sin(x):
>     elif evalf(sin(x))>0.5 then 0.5
>     else -0.5:
>     end if:
>  plot('f'(x),x=-2*Pi..2*Pi);
```

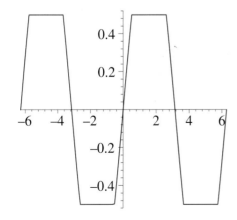

Using `piecewise`, f is defined and plotted with

```
>  f := x->piecewise(abs(sin(x))<=1/2,
   sin(x),sin(x)>1/2,1/2,-1/2):
>  plot(f(x),x=-2*Pi..2*Pi);
```

(b) Using conditionals, g is defined as follows:

```
>  g := x->
>     if x>0 and x<2 then sqrt(2*x-x^2)
>     elif x<=0 then -ln(1-x)
>     else -ln(x-1)
>     end if:
>  plot(g,-2*Pi..2*Pi);
```

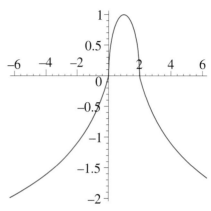

Using `piecewise`, g is defined and plotted with

```
>  g := x->piecewise(x>0 and x<2,
   sqrt(2*x-x^2),x<=0,-ln(1-x),-ln(x-1)):
>  plot(g(x),x=-2*Pi..2*Pi);
```

Solution 3.20

(a) The matrix **M** with elements M_{ij}, $i, j = 1, 2, 3$, requires a 1 on and below the diagonal ($i \geq j$), and a 0 otherwise.

```
>  M := Matrix(3,3):
>  for i from 1 to 3 do
>     for j from 1 to 3 do
>        if i>=j then M[i,j] := 1:
>        else M[i,j] := 0:  end if:
>     end do:
>  end do:
>  M;
```

$$\begin{bmatrix} 1 & 0 & 0 \\ 1 & 1 & 0 \\ 1 & 1 & 1 \end{bmatrix}$$

(b) Here, we require $M_{ij} = 1$ for $i < j$, $M_{ij} = -1$ for $i > j$ and $M_{ij} = 0$ otherwise:

```
>  M := Matrix(3,3):
>  for i from 1 to 3 do
>     for j from 1 to 3 do
>        if i<j then M[i,j] := 1:
>        elif i>j then M[i,j] := -1:
>        else M[i,j] := 0:  end if:
>     end do:
>  end do:
>  M;
```

$$\begin{bmatrix} 0 & 1 & 1 \\ -1 & 0 & 1 \\ -1 & -1 & 0 \end{bmatrix}$$

(c) Here, M_{ij} takes three possible values and we proceed as follows:

```
>   M := Matrix(3,3):
>   for i from 1 to 3 do
>     for j from 1 to 3 do
>       if i+j<=3 then M[i,j] := 1:
>       elif i+j>=5 then M[i,j] := -1:
>       else M[i,j] := 0:  end if:
>     end do:
>   end do:
>   M;
```

$$\begin{bmatrix} 1 & 1 & 0 \\ 1 & 0 & -1 \\ 0 & -1 & -1 \end{bmatrix}$$

Solution 3.21

The following is based on the solution to Exercise 3.6. The while option terminates the loop execution when $|x - N^{1/n}| < 10^{-8}$.

```
>   x := 1:  N := 3:  n := 2:
>   for k while abs(evalf(x-N^(1/n)))
    >=10^(-8) do
>     x := (N/x^(n-1)+(n-1)*x)/n:
>     print(k,evalf(x-N^(1/n))):
>   end do:
```

$$1, 0.267949192$$
$$2, 0.017949192$$
$$3, 0.000092049$$
$$4, 0.2 \ 10^{-8}$$

Note that evalf has to be used to get the conditional to work.

Solution 3.22

(a) The conditional in the second loop is modified as shown on line #7:

```
>   N := 128:  pk := 2:  pxs := NULL:        #1
>   for k while pk<=N-2 do                   #2
>     pxs := pxs,pk:                          #3
>     pk := nextprime(pk):                    #4
>   end do:                                   #5
>   for k in pxs do                           #6
>     if isprime(N-k) and k>=N-k then         #7
>       print(k,N-k):                         #8
>     end if:                                 #9
>   end do:                                   #10
```

$$67, 61$$
$$97, 31$$
$$109, 19$$

(b) In order to calculate the number of prime summands of N, we initialize a variable num to zero on line #1, and replace the print statement on line #8 with num := num+1. This variable therefore counts the number of prime pairs which sum to N.

```
>   N:=128:  pk := 2:  pxs := NULL:
    num := 0:                                 #1
>   for k while pk<=N-2 do                    #2
>     pxs := pxs,pk:                           #3
>     pk := nextprime(pk):                     #4
>   end do:                                    #5
>   for k in pxs do                            #6
>     if isprime(N-k) and k>=N-k then          #7
>       num := num+1:                          #8
>     end if:                                  #9
>   end do:                                    #10
>   num;                                       #11
```

$$3$$

Solution 3.23

(a) The following loop ensures that pxs is a sequence of all odd primes less than or equal to $N - 6$ (where $N = 25$).

```
>   N := 25:  pk := 3:  pxs := NULL:
>   for k while pk<=N-6 do
>     pxs := pxs,pk:
>     pk := nextprime(pk):
>   end do:
>   pxs;
```

$$3, 5, 7, 11, 13, 17, 19$$

(b) Here i loops over all odd primes less than $N - 6$, and j loops over all odd primes less than or equal to i. If $N - i - j$ is prime and $N - i - j \leq j$, then $i, j, N - i - j$ is printed. This ensures that $i \geq j \geq N - i - j$.

```
>   for i in pxs do
>     for j in pxs while j<=i do
>       if isprime(N-i-j) and N-i-j<=j then
>         print(i,j,N-i-j):
>       end if:
>     end do;
>   end do;
```

$$11, 7, 7$$
$$11, 11, 3$$
$$13, 7, 5$$
$$17, 5, 3$$
$$19, 3, 3$$

Solution 3.24

(a) This is quite a simple procedure:

```
>   r := proc(x,y,z)
>      sqrt(x^2+y^2+z^2):
>   end proc:
```

Test the procedure:

```
>   r(3,4,5);
```
$$5\sqrt{2}$$

(b) A good idea is to convert the code from Exercise 3.3(a) to a procedure.

```
>   S := proc(N)
>      local sm,k:
>      sm := 0:
>      for k to N do
>         sm := sm+(-1)^k/k:
>      end do:
>      evalf(sm):
>   end proc:
```

Now check that we get the same result as in Exercise 3.3(a):

```
>   S(100);
```
$$-0.6881721793$$

(c) One way is to proceed as follows:

```
>   fact := proc(n,k)              #1
>      local prod,r:               #2
>      prod := 1:                  #3
>      for r from n to 1 by -k do  #4
>         prod := prod*r:          #5
>      end do:                     #6
>      prod:                       #7
>   end proc:                      #8
```

In this procedure there are two local variables, `prod` and `r`.

The first iteration of the loop is executed if `r` (which in this iteration has its value set to `n`) is greater than or equal to 1. On this first iteration:

on line #5, `prod` is assigned the value of `n`.

The loop goes to the second iteration if `r` (which in this iteration has its value decreased by `k`) is greater than or equal to 1. On this second iteration:

on line #5, the value of `prod` is multiplied by `n-k`.

The loop goes to the third iteration if `r` (which in this iteration has its value again decreased by `k`) is greater than or equal to 1.

And so on.

Now check that the procedure reproduces the results from Exercise 3.5:

```
>   fact(10,1);
```
$$3628800$$

```
>   fact(20,3);
```
$$4188800$$

Solution 3.25

The code to plot this function is straightforward. We use `mul` to evaluate the product.

```
>   C2 := proc(N)
>      local res:
>      res :=
         mul(1-1/(ithprime(k)-1)^2,k=2..N):
>      evalf(res):
>   end proc:
```

To plot `C2` we use `seq` to create a sequence of data points $[50, C2(50)], [51, C2(51)], \ldots, [500, C2(500)]$.

```
>   dat := seq([k,C2(k)],k=50..500):
>   plot([dat]);
```

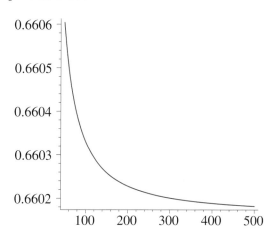

So $C2(N)$ appears to converge to a value just below 0.6602. For larger values of N we observe that $C2(N) \simeq 0.66016$:

```
>   C2(3000);
```
$$.6601639703$$

Solution 3.26

We write a procedure based on the loop in Solution 3.6:

```
>   iter := proc(N,n,M,x0)            #1
>      local x,k:                     #2
>      x := x0:                       #3
>      for k to M do:                 #4
>         x := (N/x^(n-1)+(n-1)*x)/n: #5
>      end do:                        #6
>      evalf(x):                      #7
>   end proc:                         #8
```

Note that `x0` cannot be used on the left-hand side of an assignment statement as it occurs in the argument list. So on line #3 we assign its value to the local variable `x`, and use this in the loop.

For $N = 23$, $n = 3$, $x_0 = 1$, this procedure gives

```
>   evalf(iter(23,3,6,1)-23^(1/3));
```
$$0.000297216$$

for the value of $x_6 - 23^{1/3}$, which is the same result as as in Solution 3.6(c).

Solution 3.27

The following procedure is closely based on `polar2` on page 135.

```
>   polar3 := proc(x,y,z)
>     local r,theta,phi:
>     if x=0 and y=0 and z<>0 then
>       print("phi not defined"):
>       return:
>     end if:
>     if x=0 and y=0 and z=0 then
>       print("theta and phi not defined"):
>       return:
>     end if:
>     r := sqrt(x^2+y^2+z^2):
>     phi := arctan(y,x):
>     theta := arctan(sqrt(x^2+y^2),z):
>     r,theta,phi:
>   end proc:
```

Test it:

```
>   polar3(2,-4,0);
```

$$2\sqrt{5}, \frac{\pi}{2}, -\arctan(2)$$

Solution 3.28

The following code outputs the angle, and checks that the input is of the correct data type:

```
>   angle3 := proc(v::Vector(3),w::Vector(3))
>     local ct:
>     ct := v.w/(sqrt(v.v)*sqrt(w.w)):
>     evalf(arccos(ct)):
>   end proc:
```

For example, the angle between the vectors $\mathbf{a} = \mathbf{i} + 2\mathbf{j}$ and $\mathbf{b} = \mathbf{i} + 2\mathbf{j} + 3\mathbf{k}$ is given by

```
>   a := Vector([1,2,0]):
    b := Vector([1,2,3]):
>   angle3(a,b);
```

$$0.9302740142$$

Solution 3.29

(a) The procedure is very similar to that shown in the text:

```
>   gr1 := proc(a::realcons,rnge::range,optns)
>     local eq,ic,sol,fy,x0:
>     eq := D(y)(x)=cos(Pi*x*y(x));
>     ic := y(0)=a:
>     sol := dsolve({eq,ic},numeric):
>     fy := x0->eval(y(x),sol(x0));
>     plot(fy,rnge,optns);
>   end proc:
```

(b) To plot the solutions for $a = 1, 2, 3$, over the range $0 \le x \le 6$, we use three different line colours:

```
>   plots[display]([gr1(1,0..6,colour=red),
    gr1(2,0..6,colour=black),
    gr1(3,0..6,colour=grey)]);
```

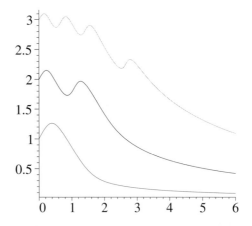

Solution 3.30

(a) We modify the code produced in Exercise 3.22(b) in a straightforward way.

```
>   numprime := proc(N::integer)                    #1
>     local pk,pxs,num,k:                            #1
>     if N<=2 or frac(N/2)<> 0 then                  #2
>       print("Error, N must be >2
              and even"):                            #3
>       return:                                      #4
>     end if:                                        #5
>     pk := 2:  pxs := NULL:  num := 0:              #6
>     for k while pk<=N-2 do                         #7
>       pxs := pxs,pk:                               #8
>       pk := nextprime(pk):                         #9
>     end do:                                        #10
>     for k in pxs do                                #11
>       if isprime(N-k) and k>=N-k then              #12
>         num := num+1:                              #13
>       end if:                                      #14
>     end do:                                        #15
>     num:                                           #16
>   end proc:                                        #17
```

We test this procedure:

```
>   numprime(128);
```

$$3$$

so there are 3 prime pairs which sum to 128, which agrees with the solution to Exercise 3.22(b).

Lines #2 to #5 simply check that N is greater than 2 and even. Lines #6 to #15 are taken straight from the solution to Exercise 3.22(b). Line #16 returns the number of prime summands of N.

(b) To plot $\nu(N)$ for $4 \le N \le 2000$, we construct a list of data points:

$$[[4, \texttt{numprime(4)}],$$
$$[6, \texttt{numprime(6)}],$$
$$\vdots$$
$$[2000, \texttt{numprime(2000)}]].$$

We do this as follows:

```
>  dat :=
   [seq([2*k,numprime(2*k)],k=2..1000)]:
>  p1 := plot(dat,style=point,colour=black):
```

The points in Figure 3.13 show the graph produced.

(c) From Exercise 3.25, the twin-prime constant is $C_2 \simeq 0.6602$.

```
>  C2 := 0.6602:
>  nump := C2*n/(ln(n))^2:
>  p2 := plot(nump,n=1..2000,color=red):
```

Now display both p1 and p2 on the same graph:

```
>  plots[display]([p1,p2]);
```

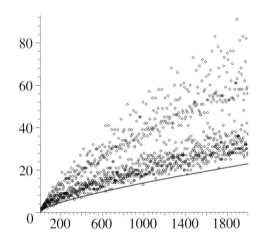

Figure 3.13

From this graph we observe that $C_2 \dfrac{N}{(\ln(N))^2}$ provides an accurate estimate of the smallest number of prime summands of N.

Solution 3.31

A good way to proceed is to define a function $f(p) = 2^p - 1$, and loop over the first 100 primes p_k. If $f(p_k)$ is prime, then p_k is printed.

```
>  f := p->2^p-1:
>  for k from 1 to 100 do
>     pk := ithprime(k):
>     if isprime(f(pk)) then print(pk):
>     end if:
>  end do:
```

The output generates the numbers 2, 3, 5, 7, 13, 17, 19, 31, 61, 89, 107, 127, 521.

Note that Mersenne's list is not correct. He mistakenly included M_{67} and M_{257}, and omitted M_{61}, M_{89} and M_{107}. Maple has found that M_{521} is also a Mersenne prime.

The search for Mersenne primes is still an active one and new ones are found quite regularly. You may like to search on the internet for the latest list, and for news of newly discovered ones.

Solution 3.32

A straightforward way to write this function is as follows.

```
>  T := x->
>     if x<1 then 1:
>     elif x>=1 and x<=2 then 2:
>     elif x>2 and x<3 then 1:
>     elif x>=3 and x<=4 then 2:
>     else 1:
>     end if:
```

The corresponding piecewise command would be

```
>  Tp := x->piecewise(x<1,1,x>=1 and x<=2,2,
      x>2 and x<3,1,x>=3 and x<=4,2,1);
```

However, some of these tests are redundant. Thus the conditional can be written more simply as

```
>  T := x->
>     if x<1 then 1:
>     elif x<=2 then 2:
>     elif x<3 then 1:
>     elif x<=4 then 2:
>     else 1:
>     end if:
```

Then the piecewise version can be written

```
>  Tp := x->piecewise(x<1,1,
         x<=2,2,x<3,1,x<=4,2,1):
```

Either way, we check our function by plotting it:

```
>  plot('T'(x),x=-2..6,0..2);
```

for the conditional, and

```
>  plot(Tp(x),x=-2..6,0..2);
```

for the piecewise defined function (see the output below).

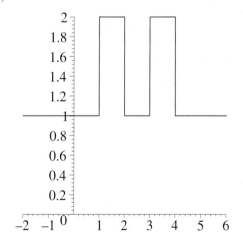

Solution 3.33

(a) In the following, we use P||n to represent the nth spherical Legendre polynomial. Note that we use simplify to obtain as 'simple' a form as possible.

```
>   restart:  N := 5:
>   P||0 := 1;  P||1 := x;
>   for n to N-1 do
>      P||(n+1) := simplify(((2*n+1)*x*P||n
                  -n*P||(n-1))/(n+1));
>   end do;
```

$$P0 := 1$$
$$P1 := x$$
$$P2 := \frac{3\,x^2}{2} - \frac{1}{2}$$
$$P3 := \frac{5}{2}\,x^3 - \frac{3}{2}\,x$$
$$P4 := \frac{35}{8}\,x^4 - \frac{15}{4}\,x^2 + \frac{3}{8}$$
$$P5 := \frac{63}{8}\,x^5 - \frac{35}{4}\,x^3 + \frac{15}{8}\,x$$

(b) Here we calculate

$$(1 - x^2)\,\frac{d^2}{dx^2}P_n - 2x\,\frac{d}{dx}P_n + n(n+1)\,P_n$$

for $n = 1, \ldots, 5$, and print out the result (after simplification):

```
>   for n to N do
>      Df := (1-x^2)*diff(P||n,x,x)
            -2*x*diff(P||n,x)+n*(n+1)*P||n:
>      print(simplify(Df));
>   end do:
```

The result is 0 in each case, showing that each polynomial satisfies the corresponding differential equation.

(c) Calculate

$$P_n(x) - \frac{1}{2^n\,n!}\,\frac{d^n}{dx^n}(x^2 - 1)^n$$

for $n = 1, \ldots, 5$:

```
>   for n to N do
>      simplify(P||n-diff((x^2-1)^n,x$n)
               /(2^n*n!));
>   end do;
```

The output is 0 in each case.

(d) Here we use combine to write the output in terms of $\cos n\theta$ ($n = 1, 2, \ldots, 5$):

```
>   for n to N do
>      combine(eval(P||n,x=cos(theta)));
>   end do;
```

$$\cos(\theta)$$
$$\frac{3}{4}\cos(2\,\theta) + \frac{1}{4}$$
$$\frac{5}{8}\cos(3\,\theta) + \frac{3}{8}\cos(\theta)$$

$$\frac{35}{64}\cos(4\,\theta) + \frac{5}{16}\cos(2\,\theta) + \frac{9}{64}$$
$$\frac{63}{128}\cos(5\,\theta) + \frac{35}{128}\cos(3\,\theta) + \frac{15}{64}\cos(\theta)$$

(e) We use a double loop to print out $i, j, \int_{-1}^{1} P_i(x)\,P_j(x)\,dx$ for $0 \le j \le i \le 5$:

```
>   for i from 0 to N do
>      for j from 0 to i do
>         print(i,j,int(P||i*P||j,x=-1..1));
>      end do;
>   end do;
```

$$0, 0, 2$$
$$1, 0, 0$$
$$1, 1, \frac{2}{3}$$
$$\vdots$$
$$5, 5, \frac{2}{11}$$

(To save space, we have shown only part of the Maple output that you should actually get.)

Solution 3.34

(a) To determine whether an integer N is the sum of two squares, we need to check for every $i, j \ge 1$ whether $N = i^2 + j^2$. The simplest way is to use a pair of nested loops, one looping over i and the other over j.

The question remains as to what bounds to put on the loops. The simplest would be to sum over $1 \le i, j \le \sqrt{N}$:

```
>   restart:  N := 10:
>   for i to evalf(sqrt(N)) do
>      for j to evalf(sqrt(N)) do
>         if N=i^2+j^2 then print(i,j): end if:
>      end do:
>   end do:
```

$$1, 3$$
$$3, 1$$

However, this results in some redundancy since the same pair of integers is found twice ($i = 3$, $j = 1$ and $i = 1$, $j = 3$ for $N = 10$). This is corrected by altering the upper bound of the inner loop so that $i \ge j$:

```
>   N := 10:
>   for i to evalf(sqrt(N)) do
>      for j to i do
>         if N=i^2+j^2 then print(i,j): end if:
>      end do:
>   end do:
```

$$3, 1$$

So $10 = 3^2 + 1^2$.

(b) We modify this loop as follows:

```
>  psum := proc(N::posint)
>    local i,j,Xs:                       #1
>    Xs := NULL:                         #2
>    for i to evalf(sqrt(N)) do          #3
>      for j to i do                     #4
>        if N=i^2+j^2 then Xs := Xs,[i,j]:
           end if:                       #5
>      end do:                           #6
>    end do:                             #7
>    Xs:                                 #8
>  end proc:                            #9
```

On line #2 we initialize the sequence Xs. In the conditional, whenever $N = i^2 + j^2$ we append $[i, j]$ to Xs. Finally, on line #8, Xs is returned.

Check that the procedure works:

```
>  psum(10);
```
$$[3, 1]$$

(c) We create a loop which looks for the first integer with four pairs of numbers whose squares sum to it:

```
>  for k do
>    tmp := psum(k):
>    if nops([tmp])>=4 then print(k,tmp):
>    break:  end if:
>  end do:
```
$$1105, [24, 23], [31, 12], [32, 9], [33, 4]$$

So
$$1105 = 24^2 + 23^2 = 31^2 + 12^2 = 32^2 + 9^2 = 33^2 + 4^2.$$

Solution 3.35

(a) The procedure per(x,f,a,b) takes a Maple function f defined over the range $[a, b]$ and returns the value $F(x)$ for any x.

```
>  per := proc(x::realcons,f::mathfunc,
          a::realcons,b::realcons)          #1
>    local P,y:
>    if b<=a then print("WARNING: upper
                 limit <= lower limit."):   #2
>    return:  end if:                       #3
>    P := b-a:                              #4
>    y := x - P*floor((x-a)/P):             #5
>    f(y):                                  #6
>  end proc:
```

On line #1 we demand that the argument f be a Maple function (which has data type mathfunc), while a, b and x are real numbers (of data type realcons).

On lines #2 and #3 we check that $b > a$. Note that the procedure will work without these restrictions, but you could get unexpected results if you make a mistake in your parameter data types.

On line #4 we define the variable P, while on line #5 we define a variable y.

Finally, on line #6 we return $f(y)$.

(b) Here we are asked to plot the periodic extension of $f(x) = x$, for $a = -1$, $b = 2$, $-5 \le x \le 5$:

```
>  f := x->x:
>  F := x->per(x,f,-1.0,2.0):
>  plot('F'(x),x=-5..5);
```

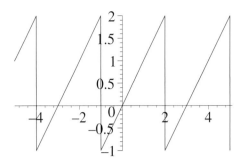

Note that we need to put quotes around the function F in order to plot it (see page 125).

Solution 3.36

(a) We modify the code produced in Exercise 3.10 to save each 3-digit sub-sequence to an expression sequence Xs:

```
>  restart:  N := 5:
   pi1 := evalf[N+1](Pi):                  #1
>  Xs := NULL:                             #2
>  for k from 1 to N-2 do                  #3
>    dig := floor(100*pi1):                #4
>    pi1 := evalf[N+1](10*frac(pi1)):      #5
>    Xs := Xs,dig:                         #6
>  end do:                                 #7
```

We have simply added lines #2 and #6 so that, with each iteration, the next 3-digit sub-sequence of π is appended to the expression sequence Xs. Hence, after three iterations of the loop, we obtain

```
>  Xs;
```
$$314, 141, 415$$

(b) The question suggests converting Xs to a set and to a list, and comparing the number of elements in each. When they differ, we must have a repeated 3-digit sub-sequence.

This is done in lines #7 and #8 below:

```
>  restart:  N := 100:
   pi1 := evalf[N+1](Pi):                  #1
>  Xs := NULL:                             #2
>  for k from 1 to N-2 do                  #3
>    dig := floor(100*pi1):                #4
>    pi1 := evalf[N+1](10*frac(pi1)):      #5
>    Xs := Xs,dig:                         #6
>    Lxs := [Xs]:  Setxs := {Xs}:          #7
>    if nops(Lxs)<>nops(Setxs) then:
       print(k,dig):  break:  end if:      #8
>  end do:                                 #9
```

When the number of elements in [Xs] differs from the number of elements in {Xs}, the loop prints the index value and the 3-digit sub-sequence.

153

The first repeated 3-digit sub-sequence in the expansion of π is obtained provided that N is assigned a large enough value in line #1. If the loop terminates without producing a result, then N should be assigned a larger value.

Here we assigned N:=100, and the code finds that the 62nd 3-digit sub-sequence (592) is repeated.

(c) We modify our code in a straightforward manner. Introduce a variable P (on line #1) to define P-digit sub-sequences. This is used on lines #2, #4 and #5 in such a way that it agrees with our code for 3- (and 1-) digit sub-sequences.

```
>   restart:  N := 1000:
    pi1 := evalf[N+1](Pi):  P := 5:        #1
>   P := P-1:                              #2
>   Xs := NULL:                            #3
>   for k from 1 to N-P do                 #4
>       dig := floor(10^P*pi1):            #5
>       pi1 := evalf[N+1](10*frac(pi1)):   #6
>       Xs := Xs,dig:                      #7
>       Lxs := [Xs]:  Setxs := {Xs}:       #8
>       if nops(Lxs)<>nops(Setxs) then:
        print(k,dig):  break:  end if:     #9
>   end do:                                #10
```

This code gives the correct results $4, 1$ on 1-digit sub-sequences, $22, 26$ on 2-digit sub-sequences, and $62, 592$ on 3-digit sub-sequences.

For P:=5 it gives the result
$$552, 60943$$
indicating that the 552nd 5-digit sub-sequence 60943 is the first to repeat.

Solution 3.37

In the following, we show simply r in our output, but you will actually see $r\sim$, because we 'assume' $r > 0$.

(a) To calculate V_2, first evaluate the inner integral:
$$\int_{-\sqrt{r^2-x_2^2}}^{\sqrt{r^2-x_2^2}} 1\, dx_1.$$

```
>   restart:  assume(r>0):
>   F0 := 1:                         #integrand
>   L1 := sqrt(r^2-x2^2):   #integration limit
>   F1 := int(F0,x1=-L1..L1);
```
$$F1: = 2\sqrt{r^2-x2^2}$$

Next the outer one: $\int_{-r}^{r} F1\, dx_2.$

```
>   L2 := sqrt(r^2):          #integration limit
>   F2 := int(F1,x2=-L2..L2);
```
$$F2: = r^2\,\pi$$

So $V_2 = \pi r^2$.

(The reason for writing $L2 = \sqrt{r^2}$ rather than $L2 = r$ will become clear in part (b).)

We follow a similar method for V_3. First calculate the inner integral with respect to x_1:

```
>   restart:  assume(r>0);
>   F0 := 1:                         #integrand
>   L1 := sqrt(r^2-x2^2-x3^2):       #limit
>   F1 := int(F0,x1=-L1..L1);
```
$$F1: = 2\sqrt{r^2-x2^2-x3^2}$$

Next integrate this with respect to x_2:

```
>   L2 := sqrt(r^2-x3^2):            #limit
>   F2 := int(F1,x2=-L2..L2);
```
$$F2: = \frac{\sqrt{r^2-x3^2}\,\pi}{\sqrt{\dfrac{1}{r^2-x3^2}}}$$

Now integrate this with respect to x_3:

```
>   L3 := sqrt(r^2):                 #limit
>   F3 := int(F2,x3=-L3..L3);
```
$$F3: = \frac{4\,r^3\,\pi}{3}$$

So $V_3 = \frac{4}{3}\pi r^3$.

(b) Now generalize the preceding code in a straightforward manner to general N (here we display the result for $N = 6$):

```
>   restart:  assume(r>0);
>   N := 6:  F||0 := 1:
>   for k from 1 to N do
>       L||k := sqrt(r^2-add(x||j^2,j=k+1..N)):
>       F||k := int(F||(k-1),x||k=-L||k..L||k);
>   end do:
>   F||N;
```
$$\frac{r^6\,\pi^3}{6}$$

So $V_6 = \pi r^6/6$.

Here the kth loop iteration calculates
$$F_k = \int_{-L_k}^{L_k} F_{k-1}\, dx_k, \quad k = 1,\ldots,N, \quad F_0 = 1,$$
where
$$L_k = \sqrt{r^2 - \sum_{j=k+1}^{N} x_j^2}.$$

Note that Maple is quite slow at calculating V_N for large N. On my computer V_6 takes about 20 seconds and V_{10} about 4 minutes.

UNIT 4   Case studies

Study guide

This unit consists of six sections, of roughly equal length. They may be studied in any order (though it makes sense to study Section 4.5 before Section 4.6). They are presented in what is felt to be a sensible order in terms of content and difficulty. Although there is material in each section that does not require a computer, you will need access to a computer at some stage in every section.

Ideally you should have time to study the whole of this unit. However, if time is short then you should try to complete at least four of the sections, including Section 4.6.

Even if you do not study Sections 4.5 and 4.6 in detail, you should introduce yourself to the `series` option in `dsolve`, which is described in Section 4.5, and to `odeplot`, which is described in Section 4.6.

Section 4.6 also uses the concatenation operation `cat` to produce plots with informative titles. Although you will not be required to use `cat` for assessment purposes, you are recommended to study this helpful facility.

Introduction

This unit looks at six case studies that explore different types of problem that can be tackled with Maple. Working through the unit should enable you to consolidate the Maple commands and structures that you learned in *Units 1–3*.

Section 4.1 uses Maple to investigate some properties of special sequences of numbers called *spectra*. You will use your computer to explore patterns that emerge from these sequences, from which conjectures about the patterns can be made. Maple can then be used to verify the conjectures for specific cases.

Section 4.2 looks at the topic of *numerical integration*. You may already be familiar with the *trapezium rule*, which can be used to approximate the value of a definite integral that cannot be found exactly. This section outlines both the trapezium rule and the more general *Newton–Cotes formula*, and shows how Maple can be used with these methods to give good approximations to definite integrals.

In Section 4.3 a method of finding polynomial approximations to functions of a single variable is introduced. Maple can be very useful here, in finding the required polynomial coefficients. The close connection between this method and Fourier series is also discussed.

Section 4.4 looks at various problems in *combinatorial mathematics*, and shows different techniques for solving them using Maple. The strategies illustrated here should be useful in a wide range of applications.

Section 4.5 explains how to find series solutions of differential equations. The method is introduced with a simple example so that the basic ideas can be understood, then Maple is used to perform the calculations that can sometimes become very unwieldy. The section concludes with a discussion on the validity of the solutions that have been calculated.

Section 4.6 looks at one particular second-order nonlinear differential equation, known as the *van der Pol equation*. This was originally formulated to describe oscillations observed in an electrical circuit, and it has no solution that can be expressed in terms of elementary functions. However, Maple can be used to explore a variety of aspects of the solutions, leading to a deeper understanding of their behaviour. This section also gives a brief introduction to *phase planes*, which are introduced formally in *Unit 1* of Block B.

Very little of the Maple work in this unit should be new to you. The exceptions are as follows: Section 4.1 uses the straightforward operations of union and subtraction on sets; Section 4.5 uses `series` as an option in `dsolve`; and in Section 4.6, `odeplot` is used to plot the solutions of differential equations, and the concatenation facility `cat` is used to create a title for a plot.

However, as stated above, the main aim of this unit is to consolidate all you have already learned, so you should find that all of the Maple code apart from the above items is familiar. The contexts within which Maple is used may well be new to you, but hopefully you will find these case studies, and in particular the applications of Maple, interesting and rewarding to study.

4.1   *Spectra of numbers*

In this section you are going to investigate special sequences called *spectra*. These turn out to have interesting properties, particularly those concerned with the way the set of positive integers can be divided into subsets, the so-called *partitioning* problems. Many of these properties, which hold for infinite sequences, have already been proved. However, Maple can be used to verify them for sequences with a large but finite number of terms. This illustrates an important application of Maple, which is to uncover patterns and conjectures that might otherwise be left undiscovered, and to verify them for particular cases. Hopefully, these will then be rigorously proved in the fullness of time.

4.1.1 Defining the spectrum of a number

The **spectrum** of a real number α, denoted $\mathrm{Spec}(\alpha)$, is defined in terms of the floor function, $\lfloor\ \rfloor$, by the following sequence:

$$\mathrm{Spec}(\alpha) = \lfloor k\alpha \rfloor, \quad k = 1, 2, 3, \ldots.$$

You may not have seen the symbol $\lfloor\ \rfloor$ before; $\lfloor x \rfloor$ is defined as the greatest integer less than or equal to x, just like the Maple `floor` function (see *Unit 3*, Subsection 3.1.3).

In *Unit 2*, Subsection 2.4.1, you saw the greatest integer less than or equal to x denoted by $[x]$; the two notations are interchangeable.

Whilst this definition holds for all real α, in this section we will consider only $\alpha > 0$. For example, the spectra of $1/3$ and 0.6 (numbers less than one) are given by

$$\begin{aligned}\mathrm{Spec}(1/3) &= 0, 0, 1, 1, 1, 2, 2, 2, 3, 3, 3, 4, 4, 4, \ldots, \\ \mathrm{Spec}(0.6) &= 0, 1, 1, 2, 3, 3, 4, 4, 5, 6, 6, 7, 7, 8, \ldots.\end{aligned} \tag{4.1}$$

You can get a feel for these by explicitly checking entries. So, for $\mathrm{Spec}(1/3)$, note that

$$\begin{aligned}\lfloor 1/3 \rfloor &= \lfloor 0.333\ldots \rfloor = 0, \\ \lfloor 2/3 \rfloor &= \lfloor 0.666\ldots \rfloor = 0, \\ \lfloor 3/3 \rfloor &= \lfloor 1 \rfloor = 1, \\ \lfloor 4/3 \rfloor &= \lfloor 1.333\ldots \rfloor = 1,\end{aligned}$$

etc., and for $\mathrm{Spec}(0.6)$,

$$\begin{aligned}\lfloor 1 \times 0.6 \rfloor &= \lfloor 0.6 \rfloor = 0, \\ \lfloor 2 \times 0.6 \rfloor &= \lfloor 1.2 \rfloor = 1, \\ \lfloor 3 \times 0.6 \rfloor &= \lfloor 1.8 \rfloor = 1, \\ \lfloor 4 \times 0.6 \rfloor &= \lfloor 2.4 \rfloor = 2,\end{aligned}$$

etc. For $4/3$ and 1.6 (numbers greater than one), we have

$$\begin{aligned}\mathrm{Spec}(4/3) &= 1, 2, 4, 5, 6, 8, 9, 10, 12, 13, 14, 16, 17, 18, \ldots, \\ \mathrm{Spec}(1.6) &= 1, 3, 4, 6, 8, 9, 11, 12, 14, 16, 17, 19, 20, 22, \ldots.\end{aligned} \tag{4.2}$$

Again, you should check the first few terms explicitly, to convince yourself of their correctness. Note that in the above examples the spectra of those numbers greater than one do not contain repetitions, whereas those for positive numbers less than one do contain repeated numbers. This is a general property of $\mathrm{Spec}(\alpha)$. For example, for $\alpha > 1$ one can see that succeeding terms in $\mathrm{Spec}(\alpha)$ are *strictly* increasing by noting that

$$\lfloor (k+1)\alpha \rfloor = \lfloor k\alpha + \alpha \rfloor > \lfloor k\alpha \rfloor,$$

where the final *strict* inequality follows precisely because $\alpha > 1$. This explains why $\mathrm{Spec}(\alpha)$ for $\alpha > 1$ contains no repetitions.

Within Maple, the spectrum of a number, for $k = 1, 2, \ldots, N$, can be determined using the following function:

```
>  spec := (x,N)->seq(floor(k*x),k=1..N):
```

In this way, one can generate the results displayed in Equations (4.1) and (4.2), but for k increased to 50, as follows:

```
>  spec(1/3,50);
```

```
0, 0, 1, 1, 1, 2, 2, 2, 3, 3, 3, 4, 4, 4, 5, 5, 5, 6, 6, 6, 7, 7, 7, 8, 8, 8,
9, 9, 9, 10, 10, 10, 11, 11, 11, 12, 12, 12, 13, 13, 13, 14, 14, 14,
15, 15, 15, 16, 16, 16
```

```
>  spec(0.6,50);
```

> 0, 1, 1, 2, 3, 3, 4, 4, 5, 6, 6, 7, 7, 8, 9, 9, 10, 10, 11, 12, 12, 13, 13,
> 14, 15, 15, 16, 16, 17, 18, 18, 19, 19, 20, 21, 21, 22, 22, 23, 24, 24,
> 25, 25, 26, 27, 27, 28, 28, 29, 30

```
>  spec(4/3,50);
```

> 1, 2, 4, 5, 6, 8, 9, 10, 12, 13, 14, 16, 17, 18, 20, 21, 22, 24, 25, 26,
> 28, 29, 30, 32, 33, 34, 36, 37, 38, 40, 41, 42, 44, 45, 46, 48, 49,
> 50, 52, 53, 54, 56, 57, 58, 60, 61, 62, 64, 65, 66

```
>  spec(1.6,50);
```

> 1, 3, 4, 6, 8, 9, 11, 12, 14, 16, 17, 19, 20, 22, 24, 25, 27, 28, 30, 32,
> 33, 35, 36, 38, 40, 41, 43, 44, 46, 48, 49, 51, 52, 54, 56, 57, 59, 60,
> 62, 64, 65, 67, 68, 70, 72, 73, 75, 76, 78, 80

For the sequence lengths of this section, the default setting of `Digits` (i.e. `Digits:=10`) will suffice, as the values in `spec(x,N)` will not get too big. However, if you explore spectra for very large `N`, you may need to increase `Digits` to a sufficiently high value in order to generate the correct spectrum.

4.1.2   Partitioning the positive integers

The examples given in Subsection 4.1.1 were for the spectra of *rational* positive numbers. Let us now consider the spectrum of an *irrational* number such as $\sqrt{2}$, where the Maple function `spec` created in the previous subsection confirms that

$$\text{Spec}(\sqrt{2}) = 1, 2, 4, 5, 7, 8, 9, 11, 12, 14, 15, 16, 18, 19, 21, 22, 24, \ldots .$$

Note that since $\sqrt{2} > 1$, numbers in this sequence do not appear more than once. Compare this sequence with the spectrum of another irrational number:

$$\text{Spec}(2 + \sqrt{2}) = 3, 6, 10, 13, 17, 20, 23, 27, 30, 34, 37, 40, 44, 47, 51, \ldots ,$$

where, again, no number occurs more than once. However, these two sequences have the following remarkable property: *any* positive integer missing in one sequence occurs in the other, and the same integer *never* appears in both. When this happens, we say that the two sequences **partition** the set of positive integers $\mathbb{Z}^+ = \{1, 2, 3, \ldots\}$, or we say that $\mathbb{Z}^+$ is the **disjoint union** of the two spectra.

It turns out that this is a particular example of the following more general result. If α_1 and α_2 are positive irrational numbers satisfying

$$\frac{1}{\alpha_1} + \frac{1}{\alpha_2} = 1, \tag{4.3}$$

then $\text{Spec}(\alpha_1)$ and $\text{Spec}(\alpha_2)$ partition $\mathbb{Z}^+$. Note that in the earlier example, $\alpha_1 = \sqrt{2}$ and $\alpha_2 = \alpha_1/(\alpha_1 - 1) = 2 + \sqrt{2}$. Rather than actually prove this result, Exercise 4.1 will use Maple to verify it for a particular subset of suitable α, for arbitrarily large sequence sizes.

If one of α_1, α_2 in Equation (4.3) is rational, then clearly so is the other. Hence (using proof by contradiction) if one is irrational, then so is the other.

Before trying the following exercises, you need to know that you can 'join' two sequences into a set by placing them in curly brackets. For example, $2, 3$ and $3, 4$ can be joined as follows:

```
>   s1 := 2,3:   s2 := 3,4:
```
```
>  {s1, s2};
```

$$\{2, 3, 4\}$$

The result is the set-theoretic union of the two sequences (regarded as sets $\{2,3\}$ and $\{3,4\}$).

You can also find the set-theoretic difference of two *sets* by using `minus`. For example:

```
>   A := {1,2,3,4,5}:  B := {1,3,4}:
>   A minus B;
```

$$\{2, 5\}$$

Before starting the next three exercises, you should begin a worksheet by defining the spectrum function as in Subsection 4.1.1. The solutions at the end of the unit assume that this has been done.

Exercise 4.1

(a) Choose some arbitrary irrational number $\alpha_1 > 1$, and let $\alpha_2 = \alpha_1/(\alpha_1 - 1)$. Then use Maple to confirm, for sequence sizes up to $N = 1000$, that no integer appears more than once in either $\mathrm{Spec}(\alpha_1)$ or $\mathrm{Spec}(\alpha_2)$, and that if an integer does occur in one spectrum, then it does not occur in the other.

(b) Show, using Maple, that the set formed by uniting the two spectra in part (a) contains all the positive integers less than some $N_{\max}$, and determine $N_{\max}$ for your particular choice of α_1 in part (a). The value of $N_{\max}$ can be made as large as desired by increasing N, thus verifying the required result for arbitrarily large sequence sizes.

Exercise 4.2

The previous exercise involved comparing $\mathrm{Spec}(\alpha)$ and $\mathrm{Spec}(\alpha/(\alpha - 1))$ for any irrational $\alpha > 1$. Now use Maple to investigate instead $\mathrm{Spec}(\alpha)$ and $\mathrm{Spec}(\alpha/(\alpha + 1))$ for some positive real values of α (rational as well as irrational), and explain how they are related.

4.1.3   The spectrum of the golden ratio

Let ϕ denote the irrational number

$$\phi = (1 + \sqrt{5})/2 \simeq 1.618\,033\,99,$$

which is called the **golden ratio**. Consider the spectra of ϕ and ϕ^2:

$$\mathrm{Spec}(\phi) = 1, 3, 4, 6, 8, 9, 11, 12, 14, 16, 17, 19, 21, 22, 24, 25, 27, 29, \ldots,$$

$$\mathrm{Spec}(\phi^2) = 2, 5, 7, 10, 13, 15, 18, 20, 23, 26, 28, 31, 34, 36, 39, 41, 44, \ldots.$$

If you have studied the course MS221, you may recall that ϕ is the positive solution of the quadratic equation
$\phi^2 - \phi - 1 = 0$.

Note that since ϕ is an irrational number greater than one, and the defining equation can be written

$$\frac{1}{\phi} + \frac{1}{\phi^2} = 1,$$

the spectra partition $\mathbb{Z}^+$. However, the spectra also have another interesting property, which is investigated in the following exercise.

Exercise 4.3

Let a_k and b_k denote the kth elements of $\mathrm{Spec}(\phi)$ and $\mathrm{Spec}(\phi^2)$, respectively. By inspecting the spectra displayed above, you will notice that a_k and b_k are related as follows:

$$b_k = k + a_k, \tag{4.4}$$

for all the given values of k.

(a) Write a Maple program which tests the relation given by Equation (4.4) for all k when $\mathrm{Spec}(\phi)$ and $\mathrm{Spec}(\phi^2)$ have 1000 elements each.

(b) By rewriting the defining equation as

$$\phi^2 = 1 + \phi,$$

prove that the relation (4.4) must hold.

4.1.4   Partitioning by nonhomogeneous spectra

Subsection 4.1.2 described how $\mathbb{Z}^+$ is partitioned by $\mathrm{Spec}(\alpha_1)$ and $\mathrm{Spec}(\alpha_2)$ provided that α_1 and α_2 satisfy Equation (4.3). But this holds *only* if α_1 and α_2 are positive *irrational* numbers. Moreover, it is known that $\mathbb{Z}^+$ *cannot* be partitioned into the disjoint union of three or more spectra. However, one can lift these restrictions by extending the definition of $\mathrm{Spec}(\alpha)$ to include two arguments. Thus, we define a new sequence by

$$\mathrm{Spec}(\alpha; \beta) = \lfloor k\alpha + \beta \rfloor, \quad k = 1, 2, 3, \ldots.$$

This is sometimes referred to as a **nonhomogeneous spectrum** (as opposed to the previous definition of $\mathrm{Spec}(\alpha)$, which is correspondingly referred to as homogeneous). Before proceeding further, you might like to investigate $\mathrm{Spec}(\alpha; \beta)$ for various α and β; for example, what are the first five terms in $\mathrm{Spec}(3; 5/2)$?

With this definition, it is possible to partition $\mathbb{Z}^+$ into two *or more* nonhomogeneous spectra of *rational* numbers in the following way. For any integer $m \geq 2$, $\mathbb{Z}^+$ can be partitioned into the disjoint union of m nonhomogeneous spectra

$$\mathrm{Spec}(\alpha_n; \beta_n), \quad n = 1, 2, \ldots, m,$$

where α_n and β_n are the rational numbers given by

$$\alpha_n = \frac{2^m - 1}{2^{n-1}}, \quad \beta_n = 1 - 2^{m-n}, \quad n = 1, 2, \ldots, m. \tag{4.5}$$

Note that

$$\sum_{n=1}^{m} \frac{1}{\alpha_n} = \frac{1}{2^m - 1} \sum_{n=1}^{m} 2^{n-1}.$$

You probably know from previous mathematical study that

$$\sum_{n=1}^{m} 2^{n-1} = 2^m - 1,$$

giving the equation

$$\sum_{n=1}^{m} \frac{1}{\alpha_n} = 1,$$

The formula for the sum of a geometric progression is in the course *Handbook*.

which is analogous to the relationship in Equation (4.3) for two irrational numbers α_1 and α_2.

Exercise 4.4

Use Maple to verify the above result (that of partitioning $\mathbb{Z}^+$ into m nonhomogeneous spectra) for sequences up to size 1000 and for

(a) $m = 2$,   (b) $m = 3$.

In each case, use the method of Exercise 4.1 to show that the set formed by uniting the m spectra contains all the positive integers less than some $N_{\max}$.

4.2   Numerical integration

Numerical integration is the name given to a range of numerical methods used for approximating definite integrals. Such numerical approximations are necessary when the integral cannot be calculated exactly; an example of such an integral is

$$A = \int_1^2 \frac{e^{-x^2}}{x+1}\, dx. \tag{4.6}$$

In Maple the command `evalf(Int(y,x=a..b))` is able to evaluate numerical integrals using a variety of sophisticated methods. You saw this construction in *Unit 1*, Subsection 1.6.2.

In this section you are going to use Maple to examine two numerical integration techniques, in order to understand how they work.

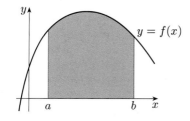

Figure 4.1

4.2.1   The trapezium rule

Recall that the definite integral

$$A = \int_a^b f(x)\, dx \tag{4.7}$$

gives the area lying under the graph of $f(x)$ for $a \le x \le b$ (see Figure 4.1). (The area is regarded as positive when $f(x)$ lies above the x-axis, and negative when it lies below the x-axis.)

One simple method of approximating this area is depicted in Figure 4.2. The area below the curve is divided into three vertical strips of equal width, and each strip is approximated by a trapezium (shaded region). The width of each trapezium is $h = (b - a)/3$, so the points on the x-axis correspond to

$$x_0 = a, \quad x_1 = a + h, \quad x_2 = a + 2h, \quad x_3 = a + 3h = b.$$

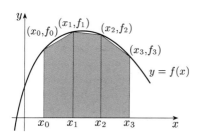

Figure 4.2 Approximating the area under the graph by trapezia; here f_i denotes $f(x_i)$ $(i = 0, 1, 2, 3)$

161

The height of the graph at each point x_i ($i = 0, 1, 2, 3$) is given by $f(x_i)$, so the area of the shaded region in Figure 4.2 is given by the sum of the areas of the three trapezia:

$$\int_a^b f(x)\, dx \simeq \frac{h}{2}(f(x_0) + f(x_1)) + \frac{h}{2}(f(x_1) + f(x_2)) + \frac{h}{2}(f(x_2) + f(x_3))$$

$$= \frac{h}{2}(f(a) + f(b)) + h(f(a + h) + f(a + 2h)), \qquad (4.8)$$

where $h = (b - a)/3$.

Exercise 4.5

(a) Use Maple to approximate the value of the definite integral in Equation (4.6) using three trapezia.

(b) Does the value calculated underestimate or overestimate the integral? Plot a graph of the integrand to explain why.

A better approximation to the area A is obtained if the number of trapezia is increased. Suppose that the area between a and b is divided into N trapezia each of width $h = (b - a)/N$. Then following an analogous argument to that given above, it can be shown that the area covered by these trapezia is

$$\int_a^b f(x)\, dx \simeq \frac{h}{2}(f(a) + f(b)) + h\sum_{k=1}^{N-1} f(a + kh), \quad h = \frac{b - a}{N}. \qquad (4.9)$$

This simple numerical integration formula is known as the **trapezium rule**. Notice that the integral has been replaced by a sum of values of the integrand calculated at points in the range of integration.

Exercise 4.6

Derive Equation (4.9).

Exercise 4.7

(a) Write a Maple procedure **trap** whose output is the approximate value of the definite integral $\int_a^b f(x)\, dx$ obtained by using the trapezium rule. The input should be the function f, the limits a and b, and the number of trapezia N.

(b) By using the function in Equation (4.6), check that your procedure correctly reproduces the result you obtained in Exercise 4.5(a).

(c) Plot a graph of $\log_{10}|A - A_N|$ versus N for $N = 3, 4, \ldots, 10$, where A is the value that Maple gives for the integral in Equation (4.6), and A_N is the approximate value of the integral given by the trapezium rule with N trapezia.

4.2.2   The Newton–Cotes formula

The trapezium rule provides accurate approximations to definite integrals provided that one takes large enough values of N (the error decreasing with increasing N). However, if the integrand is computationally time-consuming to calculate, then this method is not appropriate. Instead one requires a method which is as efficient as possible, in the sense that it requires the least number of evaluations of the integrand to achieve a given accuracy. In this section we explore a generalization of the trapezium rule known as the **Newton–Cotes formula**.

The formula is named after the English mathematicians and scientists Isaac Newton (1643–1727) and Roger Cotes (1682–1716).

The formula is normally much more efficient than the trapezium rule, is relatively easy to understand, and provides an opportunity for you to use a variety of Maple commands.

The $(N+1)$-point **closed Newton–Cotes formula** approximates a definite integral with a sum of the form

'Open' Newton–Cotes formulae also exist, but are not considered in this course.

$$\int_a^b f(x)\,dx \simeq h\sum_{j=0}^{N} c_j\, f(a + jh), \tag{4.10}$$

where $h = (b - a)/N$ and $c_0, c_1, \ldots, c_N$ are coefficients yet to be determined.

Note that the sum is over equally spaced points in the range of integration, and is clearly a generalization of the trapezium rule given by Equation (4.9).

A crucial concept associated with numerical integration is the notion of **degree of accuracy**. A numerical integration formula is said to have degree of accuracy m if it is exact for all polynomials of degree less than or equal to m (but is not exact for polynomials of degree greater than m). Later in this subsection we use this definition to calculate the coefficients c_j, in order to obtain a formula which is exact for the set of polynomials of degree less than or equal to m.

The phrase 'the set of polynomials of degree less than or equal to m' is something of a mouthful. This set is actually quite easy to specify, as the set of polynomials of the form $a_0 + a_1 x + \cdots + a_m x^m$, as $a_0, \ldots, a_m$ range independently over all real values. In the remainder of this section we shall denote this set by $\mathcal{P}_m$.

The Newton–Cotes formula for $N = 1$

We first consider the closed Newton–Cotes formula for the simple case when the lower limit is $a = 0$ and the upper limit is $b = B$. For this special case we call the integrand $g(x)$, so that Equation (4.10) becomes

$$\int_0^B g(x)\,dx \simeq h\sum_{j=0}^{N} c_j\, g(jh), \quad \text{where } h = B/N. \tag{4.11}$$

It suffices to consider this case, because the coefficients c_j don't change if the integration limits are both shifted by some constant. You are asked to prove this in the following exercise.

Exercise 4.8

(a) Show that, with an appropriate change of variable $y = x - a$ and function $g(y) = f(y + a)$,

$$\int_a^b f(x)\,dx = \int_0^B g(y)\,dy, \quad \text{where } B = b - a.$$

163

(b) Show that if

$$\int_0^B g(x)\,dx \simeq h\sum_{j=0}^{N} c_j\, g(jh), \quad \text{where } h = B/N,$$

has degree of accuracy m for coefficients c_j (with $B = b - a$), then Equation (4.10) has the same degree of accuracy for the same coefficients.

The result of Exercise 4.8 shows that knowing how to find the coefficients $c_0, c_1, \ldots, c_N$ in (4.11) is enough, in the sense that the more general problem (4.10) can easily be converted to that in (4.11), and the c_j calculated for this problem are then correct for the original problem.

As mentioned, we determine the coefficients c_j by demanding that the formula is exact for all polynomials up to as high a degree as possible. In the general case, the $N + 1$ coefficients $c_0, c_1, \ldots, c_N$ are determined by demanding that the formula is exact for all polynomials in $\mathcal{P}_N$. In the $N = 1$ case there are two coefficients, c_0 and c_1, so we require the formula to be exact for all polynomials in $\mathcal{P}_1$.

For the moment let's restrict ourselves to the simple $N = 1$ case, which we solve by hand in order to illustrate the general method. Here $h = B$, so Equation (4.11) becomes

$$\int_0^B g(x)\,dx \simeq B(c_0\, g(0) + c_1\, g(B)). \tag{4.12}$$

In order for this formula to be exact for all polynomials in $\mathcal{P}_1$, that is, all polynomials of the form $g(x) = a_0 + a_1 x$, we require that

$$\int_0^B (a_0 + a_1 x)\,dx = B(c_0 a_0 + c_1(a_0 + a_1 B)), \tag{4.13}$$

for all a_0 and a_1. Integrating the left-hand side, and collecting coefficients of a_0 and a_1 on the right-hand side, gives

$$a_0 B + \tfrac{1}{2} a_1 B^2 = B(c_0 + c_1)a_0 + B^2 c_1 a_1. \tag{4.14}$$

Since a_0 and a_1 are arbitrary, we can set each of them to zero separately, to obtain two equations:

$$B = B(c_0 + c_1) \quad \text{and} \quad B^2/2 = B^2 c_1.$$

You may have seen this technique described as 'equating coefficients of a_0 and a_1'.

Since B is not zero, these have the solution

$$c_0 = c_1 = \tfrac{1}{2}.$$

Hence Equation (4.12) becomes

$$\int_0^B g(x)\,dx \simeq \frac{B}{2}(g(0) + g(B)). \tag{4.15}$$

Using the result from Exercise 4.8(b), the Newton–Cotes formula with arbitrary limits of integration becomes

$$\int_a^b f(x)\,dx \simeq \frac{h}{2}(f(a) + f(b)), \quad \text{where } h = b - a. \tag{4.16}$$

For larger values of N, the principle is the same, except that there are $N + 1$ simultaneous equations to be solved. Clearly it is advantageous to use Maple to do this.

Note that the $N = 1$ Newton–Cotes formula is simply the trapezium rule with a single trapezium. For larger values of N, the two formulae always differ.

Exercise 4.9

Use the $N = 1$ Newton–Cotes formula given by Equation (4.16) to obtain a numerical estimate for the integral in Equation (4.6).

Exercise 4.10

(a) Derive by hand the $N = 2$ Newton–Cotes formula, namely

$$\int_a^b f(x)\,dx \simeq \frac{h}{3}(f(a) + 4f(a+h) + f(b)), \quad \text{where } h = (b-a)/2.$$

(b) What value does this formula give for the integral in Equation (4.6)?

The Newton–Cotes formula using Maple

In this subsection we repeat the algebraic manipulations performed in the previous section, but this time using Maple. It will then be straightforward to generalize this code to calculate the coefficients for larger values of N.

As previously, this calculation is performed for $N = 1$ initially, but we will keep N as an arbitrary parameter so that it can be readily generalized to take other values.

First define N, h, and the first-order polynomial $p(x) = a_0 + a_1 x$:

```
>   restart:   N := 1:   h := B/N:
```

```
>   p := x->add(a[j]*x^j,j=0..N);
```

$$p := x \rightarrow \text{add}(a_j\, x^j,\ j = 0..N)$$

Now substitute this polynomial into both sides of Equation (4.11), assigning the left- and right-hand sides to the variables LH and RH, respectively:

Note that p is defined in such a way that whatever (positive integer) value we assign to N, p will be a polynomial in $\mathcal{P}_N$.

```
>   LH := int(p(x),x=0..B);
```

$$LH := a_0\, B + \frac{1}{2}\, a_1\, B^2$$

```
>   RH := h*add(c[k]*p(k*h),k=0..N);
```

$$RH := B\,(c_0\, a_0 + c_1\,(a_0 + a_1\, B))$$

Reconstruct the equation by taking the difference between the two sides:

```
>   eq := RH-LH;
```

$$eq := B\,(c_0\, a_0 + c_1\,(a_0 + a_1\, B)) - a_0\, B - \frac{1}{2}\, a_1\, B^2$$

We now need to collect the coefficients of a[0] and a[1]. This is performed using the **coeff** command introduced in *Unit 1*; coeff(eq,a[0]) gives the coefficient of a[0], while coeff(eq,a[1]) gives the coefficient of a[1]. We can collect these into an expression sequence using seq:

```
>   seqeq := seq(coeff(eq,a[k])=0,k=0..N);
```

$$seqeq := B\,(c_0 + c_1) - B = 0,\ B^2\, c_1 - \frac{B^2}{2} = 0$$

To solve these equations for the coefficients c_k, first construct a sequence of the coefficients,

```
>   seqco := seq(c[k],k=0..N);
```
$$seqco := c_0, \; c_1$$

and then solve the equations `seqeq` for the unknowns `seqco`:

```
>   solve({seqeq},{seqco});
```
$$\{c_1 = \frac{1}{2}, \; c_0 = \frac{1}{2}\}$$

This clearly agrees with the values for c_0 and c_1 found earlier.

Exercise 4.11

Use the code presented above to obtain the coefficients for the $N = 2$ Newton–Cotes formula. Check your result against the formula given in Exercise 4.10(a).

Exercise 4.12

(a) Use the code above to construct a procedure NCcoef, which takes as input an integer N, and outputs a list of the coefficients $c_0, c_1, \ldots, c_N$. Check that your procedure gives the correct coefficients for the $N = 1$ and $N = 2$ cases.

(b) By modifying NCcoef, create a procedure NCint with input the function f, the limits a and b, and the integer N, and with output the value of the Newton–Cotes approximation with $N + 1$ points. Check that your procedure agrees with your earlier results.

(c) Plot a graph of $\log_{10}|A - A_N|$ versus N for $N = 3, 4, \ldots, 10$, where A is the value that Maple gives for the integral in Equation (4.6), and A_N is the approximation to this integral found using the closed Newton–Cotes formula with $N + 1$ points.

4.3   The least squares method of approximating functions

In this section we show how to generate useful approximations to functions of a single variable, on a given interval. In the first subsection we concentrate on the particular function $f(x) = \ln(1 + x)$ on the interval $[0, 1]$, in order to illustrate the essential ideas of the method.

One familiar type of approximation is the truncated **Taylor series**. The expansion of $f(x)$ about $x = 0$ is

$$F_T(x) = f(0) + xf'(0) + \frac{1}{2}x^2 f''(0) + \cdots + \frac{x^n}{n!}f^{(n)}(0) + O(x^{n+1})$$

The 'O' order notation was discussed in *Unit 2*.

$$= x - \frac{x^2}{2} + \frac{x^3}{3} - \cdots \quad \text{for } f(x) = \ln(1 + x).$$

We shall use $t_n(x)$ to refer to the approximating polynomial itself,

$$t_n(x) = f(0) + xf'(0) + \cdots + \frac{x^n}{n!}f^{(n)}(0).$$

The polynomial $t_n(x)$ is known as the *n*th-order **approximation**; the neglected terms are described as $O(x^{n+1})$.

This type of approximation is useful if $|x| \ll 1$, but as $|x|$ increases, the accuracy decreases; for $\ln(1 + x)$ the series is valid only if $|x| < 1$. The fundamental problem with this approximation is that it is generated by expanding about a single point (here $x = 0$) by using the derivatives of the function at this point, emphasizing the importance of the point of expansion over every other point. The least squares method is different and gives equal weight to every point in the interval of interest. In Subsection 4.3.1 the method is introduced using a polynomial as an approximating function.

The symbol '$\ll$' means 'very much less than'.

4.3.1 The least squares method

In the simplest implementation, the idea is that, as with Taylor approximation, we approximate the function in terms of a polynomial. Thus for a **second-order approximation** on the interval $[0, 1]$, we write

$$\ln(1 + x) \simeq L_2(x) = a_0 + a_1 x + a_2 x^2, \tag{4.17}$$

where the three coefficients are, at present, unknown.

The next step is to form $(f(x) - L_2(x))^2$, the square of the difference between the original function and the approximation. Integrate this over the interval $[0, 1]$ to construct the function

$$E(a_0, a_1, a_2) = \int_0^1 \left(f(x) - a_0 - a_1 x - a_2 x^2 \right)^2 dx. \tag{4.18}$$

This is a non-negative function of the three coefficients, and is zero only if $L_2(x) = f(x)$. The required approximation is constructed by choosing the coefficients to make $E(a_0, a_1, a_2)$ a minimum. This approximation is named the **least squares fit**. It can be shown that $E(a_0, a_1, a_2)$ has at most one stationary point and that this is the required minimum. We do not prove this here; instead we concentrate on finding the stationary point.

If you have studied the course MST209, then you will have seen the term *least squares* before. In that course only the least squares straight line through a set of discrete data points is considered.

One way to find this minimum is to use Maple to evaluate the integral in Equation (4.18) symbolically. The partial derivatives $\partial E/\partial a_k$, for $k = 0, 1$ and 2, can then be found, and the three equations $\partial E/\partial a_k = 0$ solved. The second method is to *first* partially differentiate the integral in Equation (4.18) with respect to the three variables a_0, a_1 and a_2, and *then* use Maple to evaluate the resulting expressions. The second method is preferable because it turns out to be easier to implement.

Start by differentiating the right-hand side of Equation (4.18) with respect to a_0:

$$\frac{\partial E}{\partial a_0} = \frac{\partial}{\partial a_0} \int_0^1 \left(f(x) - a_0 - a_1 x - a_2 x^2 \right)^2 dx \tag{4.19}$$

$$= \int_0^1 \frac{\partial}{\partial a_0} \left(f(x) - a_0 - a_1 x - a_2 x^2 \right)^2 dx \tag{4.20}$$

$$= -2 \int_0^1 \left(f(x) - a_0 - a_1 x - a_2 x^2 \right) dx. \tag{4.21}$$

In passing from Equation (4.19) to Equation (4.20) we have used the fact that the order of differentiation and integration can be interchanged. This is an application of the general result that

$$\text{if} \quad G(z) = \int_a^b g(x, z)\, dx, \quad \text{then} \quad \frac{dG}{dz} = \int_a^b \frac{\partial g}{\partial z}\, dx,$$

provided that both integrals exist and that a and b do not depend on z.

Similarly, the derivatives of E with respect to a_1 and a_2 are

$$\frac{\partial E}{\partial a_1} = \int_0^1 \frac{\partial}{\partial a_1} \left(f(x) - a_0 - a_1 x - a_2 x^2 \right)^2 dx$$

$$= -2 \int_0^1 x \left(f(x) - a_0 - a_1 x - a_2 x^2 \right) dx \qquad (4.22)$$

and

$$\frac{\partial E}{\partial a_2} = \int_0^1 \frac{\partial}{\partial a_2} \left(f(x) - a_0 - a_1 x - a_2 x^2 \right)^2 dx$$

$$= -2 \int_0^1 x^2 \left(f(x) - a_0 - a_1 x - a_2 x^2 \right) dx. \qquad (4.23)$$

Equations (4.21)–(4.23) can be written in the succinct form

$$\frac{\partial E}{\partial a_k} = -2 \int_0^1 \left(x^k f(x) - x^k \left(a_0 + a_1 x + a_2 x^2 \right) \right) dx, \quad k = 0, 1, 2.$$

At a stationary point of E, all the partial derivatives are zero. Hence the stationary point is given by the solution of the three equations

$$\int_0^1 x^k f(x) \, dx = \int_0^1 \left(a_0 + a_1 x + a_2 x^2 \right) x^k \, dx$$

$$= \frac{a_0}{k+1} + \frac{a_1}{k+2} + \frac{a_2}{k+3}, \quad k = 0, 1, 2. \qquad (4.24)$$

These are three linear equations for the three parameters a_0, a_1 and a_2. They generalize in a straightforward way for higher-order least squares approximations; you are asked to do this in a later exercise.

Using Maple it becomes a relatively simple matter to form and solve these equations, and the following code shows how this can be done. We have suppressed some of the output in order to save space; you should type these commands into your own computer and make sure you understand the output.

```
>   restart:
```

The function to be approximated is best defined as an expression, as follows:

```
>   f := ln(1+x):
```

The following commands are written so that they can easily be generalized to create higher-order approximations. If the second-order least squares fit defined by Equation (4.17) is required, we set

The Maple function `f := x->ln(1+x):` could be used, but then `f` must be replaced by `f(x)` when it occurs in the subsequent code.

```
>   N := 2:
>   L2 := add(a[k]*x^k,k=0..N);
```

$$L2 := a_0 + a_1 x + a_2 x^2$$

Equations (4.24) can be formed using a `loop` as follows:

```
>   for k from 0 to N do
>       eq[k] := int(x^k*f,x=0..1)=add(a[j]/(k+j+1),j=0..N);
>   end do;
```

$$eq_0 := -1 + 2\ln(2) = a_0 + \frac{1}{2} a_1 + \frac{1}{3} a_2$$

$$eq_1 := \frac{1}{4} = \frac{1}{2} a_0 + \frac{1}{3} a_1 + \frac{1}{4} a_2$$

$$eq_2 := -\frac{5}{18} + \frac{2}{3} \ln(2) = \frac{1}{3} a_0 + \frac{1}{4} a_1 + \frac{1}{5} a_2$$

The next step is to convert the coefficients in these equations to floating-point numbers. This is necessary because, although `solve` will solve the above equations symbolically to give exact solutions, if N is increased, this computation produces unwieldy expressions which are more convenient to use when converted to floating-point numbers. Moreover, if N is large (greater than 7 in this case) then the conversion of the symbolic solutions into floating-point form requires the subtraction of large numbers that erroneously give zero if the default accuracy is used, so evaluation of these expressions requires `Digits` to be increased. This problem is alleviated if the equations are converted to floating-point form before being solved. However, for large N the accuracy of the solution of the linear equations for the coefficients should be checked by increasing `Digits`.

First form a set of the $N + 1$ equations and a set of variables as follows:

```
>  eq := {seq(eq[k],k=0..N)}:  var := {seq(a[k],k=0..N)}:
```

We have suppressed the output here, to save space. To see the effect of these commands, replace the silent terminator by the noisy one.

Now convert the coefficients of these equations to floating-point numbers by typing

```
>  eq := evalf(eq):
```

and solve these equations:

```
>  sol := solve(eq,var);
```

$$sol := \{a_0 = 0.006259525837,\ a_2 = -0.2335075077,\ a_1 = 0.9157413421\}$$

Finally, substitute these values into the approximation L2 defined above:

```
>  L2 := eval(L2,sol);
```

$$L2 := 0.006259525837 + 0.9157413421\,x - 0.2335075077\,x^2$$

This is the second-order least squares fit, $L_2(x)$. Notice that simply by changing the value of N, a higher-order approximation can be obtained.

When comparing this with the original function $f(x)$, it is also interesting to compare $t_2(x)$, the second-order Taylor series of $f(x)$. This is given by

```
>  t2 := convert(series(f,x=0,N+1),polynom);
```

$$t2 := x - \frac{1}{2}\,x^2$$

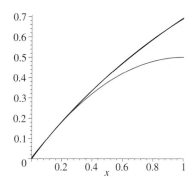

Figure 4.3 Graphs of $\ln(1 + x)$ (black), $L_2(x)$ (red) and $t_2(x)$ (red)

Figure 4.3 depicts the three functions $\ln(1 + x)$ (black), $L_2(x)$ (red, very close to $\ln(1 + x)$) and $t_2(x)$ (also red). However, this is not a very illuminating figure, because in this case the least squares fit is too close to the original function for any differences to be seen.

The inaccuracies are seen more clearly by looking at the differences

$$d_1(x) = \ln(1 + x) - t_2(x) \quad \text{and} \quad d_2(x) = \ln(1 + x) - L_2(x),$$

but it is clearer to graph the *logarithms* of the *absolute values* of $d_1(x)$ and $d_2(x)$, because this shows small differences better.

Here we use logarithms to base 10, because the numerical values along the y-axis are easier to understand, and we plot $\log_{10}(|d_k(x)| + \varepsilon)$, $k = 1, 2$, with $\varepsilon = 10^{-10}$, to avoid numerical problems should the plot procedure pick out a value of x at which $d_1(x) = 0$. These differences are compared in Figure 4.4.

Since, in this case, $L_2(0) \neq 0$ it is clear that very close to $x = 0$ the Taylor approximation (red line) is superior. However, in Figure 4.4 we see that this superiority is maintained only for $x < 0.2$, approximately, and that for $x \simeq 1$ the least squares fit is considerably superior. This improved accuracy is due to the measure of the difference, E, between $f(x)$ and $L_2(x)$ (as defined by Equation (4.18)) taking into account all relevant values of x.

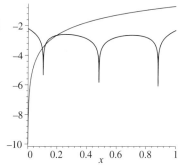

Figure 4.4 Graphs of $\log_{10}(|d_1(x)| + \varepsilon)$ (red) and $\log_{10}(|d_2(x)| + \varepsilon)$ (black)

By considering the graphs of $d_1(x)$ and $d_2(x)$ (not shown here), it can be shown that

$$|\ln(1+x) - L_2(x)| < 0.0063 \quad \text{for } 0 \le x \le 1,$$

with the largest difference being at $x = 0$, and that

$$|\ln(1+x) - t_2(x)| < 0.0063 \quad \text{only for } 0 \le x \le 0.28.$$

Exercise 4.13

Consider the nth-order Taylor series $t_n(x)$ of $\ln(1+x)$ about $x = 0$. Write a Maple procedure to evaluate $t_n(x)$ at $x = 1$. Hence find the smallest value of n for which $|t_n(1) - \ln(2)| < 0.0063$.

Exercise 4.14

(a) Use the Maple commands given in the text preceding Exercise 4.13 with $f(x) = e^x$ and $N = 3$ to show that, on $[0, 1]$, the third-order least squares fit gives

$$e^x \simeq 0.999\,060 + 1.018\,30x + 0.421\,240x^2 + 0.278\,630x^3.$$

(b) Show also that on this interval the approximation differs from the exact function by at most 0.0011.

Exercise 4.15

(a) Show that the $N + 1$ coefficients $a_0, a_1, \ldots, a_N$ of the Nth-order approximation

$$L_N(x) = a_0 + a_1 x + a_2 x^2 + \cdots + a_N x^N$$

to the function $f(x)$, on $[0, 1]$, are given by the solution of the $N + 1$ equations

$$\int_0^1 x^k f(x)\, dx = \frac{a_0}{k+1} + \frac{a_1}{k+2} + \frac{a_2}{k+3} + \cdots + \frac{a_N}{k+N+1},$$

for $k = 0, 1, 2, \ldots, N$.

(b) If this fit is required on the interval $[c, d]$, show that the equations for the coefficients found in part (a) become

$$\int_c^d x^k f(x)\, dx = \sum_{j=0}^{N} \frac{a_j}{k+j+1} \left(d^{k+j+1} - c^{k+j+1} \right),$$

for $k = 0, 1, 2, \ldots, N$.

Exercise 4.16

(a) Write a procedure LS with arguments f, interval, N and nd to find an Nth-order least squares fit to the function $f(x)$, represented by the Maple expression f, on the interval $[c, d]$, where interval is the list [c,d] and nd is the number of significant figures used in the final expression.

(b) Use your procedure to show that the 4th-order approximations to $\sin x$ and $\cos x$ on $[0, \pi]$ are, respectively,

$$\sin x \simeq 0.001\,313\,46 + 0.982\,601 x + 0.054\,469\,7 x^2 - 0.233\,793 x^3$$
$$+ 0.037\,209\,3 x^4$$

and

$$\cos x \simeq 0.990\,963 + 0.084\,984\,8 x - 0.683\,588 x^2 + 0.145\,062 x^3$$
$$+ 0.312\,166 \times 10^{-12} x^4.$$

In each case, plot the graph of the difference between the exact and approximate functions, and use this to estimate the maximum difference.

(c) Show that the maximum differences for the 8th-order approximations are about 1.3×10^{-7} and 1.6×10^{-6} for $\sin x$ and $\cos x$, respectively. In this calculation you should try the effect of changing `Digits` to 15 and 20.

4.3.2 Fourier series (optional)

In this subsection we briefly show how a simple generalization of the preceding analysis leads to the development of Fourier series.

The least squares approximation described above used a linear combination of the functions

$$\phi_0(x) = 1, \quad \phi_1(x) = x, \quad \ldots, \quad \phi_N(x) = x^N$$

to approximate a given function $f(x)$, by choosing the coefficients $a_0, a_1, \ldots, a_N$ to minimize the function

$$E(a_0, a_1, \ldots, a_N) = \int_a^b (f(x) - a_0\,\phi_0(x) - a_1\,\phi_1(x) - \cdots - a_N\,\phi_N(x))^2\,dx.$$

In the preceding analysis we chose $\phi_k(x)$ to be x^k for convenience and because these functions often provide useful approximations.

There are, however, a wide variety of suitable choices for the $\phi_k(x)$ which provide good, alternative approximations. Here we concentrate on one set, the trigonometric functions

$$1, \quad \cos x, \quad \sin x, \quad \cos 2x, \quad \sin 2x, \quad \ldots, \quad \cos Nx, \quad \sin Nx,$$

and on the interval $[-\pi, \pi]$. There are now $2N + 1$ coefficients, and the approximate function will be of the form

$$F_N(x) = a_0 + a_1 \cos x + a_2 \cos 2x + \cdots + a_N \cos Nx$$
$$+ b_1 \sin x + b_2 \sin 2x + \cdots + b_N \sin Nx.$$

A linear combination of trigonometric functions like this is sometimes called a trigonometric polynomial.

The difference function $E(a_0, a_1, \ldots, a_N, b_1, \ldots, b_N)$ is now

$$E = \int_{-\pi}^{\pi} (f(x) - F_N(x))^2\,dx,$$

and this is stationary when the following $2N + 1$ equations are satisfied:

$$0 = \frac{\partial E}{\partial a_0} = -2 \int_{-\pi}^{\pi} (f(x) - F_N(x))\,dx, \tag{4.25}$$

$$0 = \frac{\partial E}{\partial a_k} = -2 \int_{-\pi}^{\pi} (f(x) - F_N(x)) \cos kx\,dx, \quad k = 1, 2, \ldots, N, \tag{4.26}$$

$$0 = \frac{\partial E}{\partial b_k} = -2 \int_{-\pi}^{\pi} (f(x) - F_N(x)) \sin kx\,dx, \quad k = 1, 2, \ldots, N. \tag{4.27}$$

These equations simplify a great deal because, for all integers k and j,

$$\int_{-\pi}^{\pi} \cos kx \, \sin jx \, dx = 0$$

and

$$\int_{-\pi}^{\pi} \cos kx \, \cos jx \, dx = \int_{-\pi}^{\pi} \sin kx \, \sin jx \, dx = \begin{cases} 0, & k \neq j, \\ \pi, & k = j. \end{cases}$$

Thus

$$\int_{-\pi}^{\pi} F_N(x) \, dx = \int_{-\pi}^{\pi} a_0 \, dx = 2\pi a_0,$$

so Equation (4.25) gives

$$a_0 = \frac{1}{2\pi} \int_{-\pi}^{\pi} f(x) \, dx,$$

and a_0 is just the mean value of $f(x)$ on $[-\pi, \pi]$.

Similarly, Equations (4.26) and (4.27) show that

$$a_k = \frac{1}{\pi} \int_{-\pi}^{\pi} f(x) \cos kx \, dx \quad \text{and} \quad b_k = \frac{1}{\pi} \int_{-\pi}^{\pi} f(x) \sin kx \, dx,$$

for $k = 1, 2, \dots, N$.

These are the familiar equations for the **Fourier coefficients** of $f(x)$. Thus a truncated *Fourier* series for a function is a least squares fit by a *trigonometric polynomial*, in analogy with a least squares fit by an *ordinary* polynomial as found by the method of Subsection 4.3.1.

4.4   Combinatorial problem solving

The aim of this section is to illustrate some general problem-solving techniques by looking at some problems in **combinatorial mathematics**. This is the branch of mathematics concerned with counting, combining and arranging given objects in a prescribed way. In particular, we show how Maple can assist in the solution of such problems. The first subsection looks at building solutions stepwise, where the solution to a given problem is built up from the solutions to one or more smaller problems. The second subsection illustrates the 'generate and test' strategy, where the answer is obtained by selecting elements from a list or set. The third subsection looks at the mathematical basis for a magician's trick.

Several of the exercises in this section can be solved using more than one technique, and alternative methods are given in the solutions at the end of this unit. Once you have answered each exercise yourself, you are therefore advised to read through the solution, even if you have found the exercise straightforward.

4.4.1 Building solutions stepwise

Consider the problem of constructing a list of all possible orderings of n distinct objects. The usual name for a specific ordering of objects is a **permutation**, and the distinct objects are usually referred to as letters. Paradoxically, 'letters' are usually represented by numbers in modern notation, so the set of two letters is usually $\{1, 2\}$, and the list of all permutations of two letters can be represented by $[[1, 2], [2, 1]]$ in Maple. The aim of this subsection is to show how to write a procedure `permute` whose argument n is a positive integer and which returns the list of all permutations of n letters, so that typing `permute(2);` produces the output $[[1, 2], [2, 1]]$ (or, equivalently in this context, $[[2, 1], [1, 2]]$).

Start by considering the case $n = 1$, i.e. constructing a list of all the permutations of the set $\{1\}$. The only permutation of one letter is $[1]$, so a complete list of permutations is $[[1]]$.

Now consider how to proceed from this list to the case $n = 2$, i.e. the list of all permutations of $\{1, 2\}$. One method for achieving this is to insert 2 into every position in the list $[1]$ – putting the 2 into the first position yields $[2, 1]$, and putting the 2 into the second position yields $[1, 2]$. So the complete list of permutations of two letters is $[[2, 1], [1, 2]]$.

Now try to achieve this in Maple by attempting the following exercises.

Exercise 4.17

Construct a procedure `insert` that takes two arguments. The first argument is a list with N elements, and the second argument is an integer, n, with $1 \leq n \leq N + 1$. The output should be the original list with $N + 1$ inserted in the nth position. Your procedure should be able to reproduce the following:

```
>   insert([1,2,3],3);
```

$$[1, 2, 4, 3]$$

Exercise 4.18

Use the procedure `insert` defined in Exercise 4.17 to write a procedure `insertAll` whose one argument is a list L with N elements. The result should be the expression sequence of the $N + 1$ lists formed by inserting $N + 1$ into every possible position of the list L, as shown in the following:

```
>   insertAll([1]);
```

$$[2, 1], [1, 2]$$

```
>   insertAll([1,2]);
```

$$[3, 1, 2], [1, 3, 2], [1, 2, 3]$$

Now return to the problem of finding the list of all permutations of n objects. The case $n = 3$ can be solved by inserting a 3 into every position of each element of the list $[[2, 1], [1, 2]]$ in turn, as follows. Placing a 3 into the three possible positions of the first element of this list, $[2, 1]$, yields $[3, 2, 1]$, $[2, 3, 1]$ and $[2, 1, 3]$. Similarly, placing 3 into all three possible positions of the second element of the list gives $[3, 1, 2]$, $[1, 3, 2]$ and $[1, 2, 3]$. So the six possible permutations of three letters are given by the list

$$[[3, 2, 1],\ [2, 3, 1],\ [2, 1, 3],\ [3, 1, 2],\ [1, 3, 2],\ [1, 2, 3]]$$

We expect six permutations in total when $n = 3$, as $3! = 6$.

This list can be obtained by typing `map(insertAll,[[2,1],[1,2]]);`. This outputs a list containing the result of `insertAll([2,1])` followed by that of `insertAll([1,2])`.

map was introduced in Unit 2.

This is the general step: to proceed from a complete list of permutations of n letters to a complete list of permutations of $n + 1$ letters, all we need do is insert $n + 1$ into every possible position of every member of the list of permutations of n letters. Now implement this in Maple.

Exercise 4.19

Define a procedure `permute` that takes one argument n and returns the list of all permutations of n letters.

[*Hint*: Start with the list of all permutations of one letter `L:=[[1]]`, and repeatedly use `map(insertAll,L)`, with `insertAll` as defined in Exercise 4.18.]

Maple has a built-in command `permute` in the package `combinat` that achieves the same result as that requested in this exercise.

4.4.2   Generate and test

This subsection describes a strategy for solving problems that involves generating a large list and then testing the elements to find those with a desired property. The method will be illustrated by considering the solution to a famous old problem, namely the **Bernoulli–Euler problem** of the misaddressed letters:

This problem was first addressed by Nicolaus Bernoulli (1687–1759) and then later considered by Leonhard Euler (1707–1783).

> Someone writes n letters and writes the corresponding distinct addresses on n envelopes. How many different ways are there of placing all the letters in the wrong envelopes?

Both Bernoulli and Euler gave an analytic solution to this problem. Here we treat the problem as an investigation topic and show how Maple can be used to compute the number of misaddressed letters.

Consider the case $n = 3$ for ease of presentation (the extension to general n is obvious), as shown in Figure 4.5. Number the envelopes 1, 2 and 3, and consider these to be fixed. Number the letters 1, 2 and 3, and imagine permuting the letters (the permutation $[3, 2, 1]$ is shown in Figure 4.5). A permutation $L = [L_1, L_2, L_3]$ corresponds to all the letters being in the wrong envelopes if $L_i \neq i$ for $i = 1, 2, 3$. The permutation shown in Figure 4.5 has $L_2 = 2$ and so does not correspond to all the letters being misaddressed.

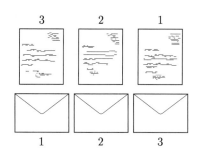

Figure 4.5   Three letters and three envelopes

In the following, you will need to use `permute` as defined in Exercise 4.19, or type `with(combinat):` to use the built-in command `permute`. Start by creating a list, Lp, of all the permutations of three letters:

If you use `with(combinat)`, you will receive a warning about a protected name, which you can safely ignore.

```
>   Lp := permute(3);
```

$$Lp := [[1, 2, 3], [1, 3, 2], [2, 1, 3], [2, 3, 1], [3, 1, 2], [3, 2, 1]]$$

You now need to find the permutations corresponding to all the letters being misaddressed. The next exercise leads you through this.

Exercise 4.20

(a) Write a procedure named `misaddressed` that returns *false* if $L_i = i$ for any i (this corresponds to a letter being in the correct envelope). Only after all the letters have been tested and found to be misaddressed should the procedure return *true*. (You have not seen how to make a procedure 'return false' or 'return true' before, but the process is very straightforward; just guess and you may well be right!)

(b) Use your procedure `misaddressed` and the `select` command to find the elements in the list `Lp` defined above that correspond to all letters being misaddressed.

`select` was introduced in Unit 2.

(c) Calculate the number of ways of misaddressing n letters, $M(n)$, for n from 1 to 8. Compare this with the corresponding result derived by Bernoulli and Euler:

$$n! \sum_{i=2}^{n} \frac{(-1)^i}{i!}.$$

4.4.3 A card trick

The following problem arises from a magician's card trick. A **perfect shuffle** occurs when a pack of cards is split exactly into two halves which are then interleaved alternately to reorder the pack. For example, consider a pack of six cards labelled 1, 2, 3, 4, 5, 6. After one perfect shuffle, the new order would be 1, 4, 2, 5, 3, 6.

Now consider a standard pack of 52 playing cards. How many perfect shuffles are needed for the pack to return to its original order? Many people believe that it would take an extraordinarily large number of shuffles to return to the original order. The following exercise shows that this is not the case, and this is the basis of the magician's card trick.

Exercise 4.21

Consider numbering a standard 52-card pack of playing cards so that the numbers from 1 to 52 appear in order. Treat this as a permutation of 52 letters, and represent it in Maple by the expression sequence `seq(i,i=1..52)`.

If you find this exercise hard, try working out by hand what happens to a pack of six cards, numbered from 1 to 6, undergoing successive perfect shuffles.

(a) Construct an expression sequence `shuffle` that corresponds to the order of the cards after one perfect shuffle.

(b) Use a `while` loop to determine the total number of perfect shuffles needed to return the pack to its original order.

[*Hint*: If the current order of the pack is the expression sequence `p`, then the order after one perfect shuffle is
`seq(p[shuffle[i]],i=1..52)`, for `shuffle` as defined above.]

Exercise 4.22

Write a procedure to calculate the number of perfect shuffles required to return to the original order, $N(n)$, for a pack with n cards (n even), and hence plot a graph of $N(n)$ for even n from 2 to 100.

[*Hint*: Use the solution to Exercise 4.21 as the basis for a procedure.]

4.5 Series solutions of differential equations

In this section we show how Maple can be used to construct the Taylor series for the solution of any first-order equation $y' = F_1(x, y)$, second-order equation $y'' = F_2(x, y, y')$, or higher, with given initial conditions. The method holds provided that the functions $F_1(x, u)$ and $F_2(x, u, v)$ are infinitely differentiable (that is, they can be differentiated as often as desired) in all variables at the initial value of the independent variable (denoted by x in this section). This type of approximation is sometimes useful because most differential equations cannot be solved in terms of known functions. In these circumstances, the Taylor series provides a general solution, although usually one that is valid only in a limited range of the independent variable x. This method provides a complementary approximation to that given by numerical techniques, which provide solutions for only particular initial conditions, though usually for a larger range of x.

4.5.1 The general method

The truncated Taylor series can be derived by hand, but the algebra quickly becomes unwieldy. However, it is important that you understand the process, so we first illustrate the method with two examples. Then we show how Maple can perform the task effortlessly.

Example 4.1

Consider the simple first-order equation

$$\frac{dy}{dx} = y, \tag{4.28}$$

with initial condition $y(0) = 1$, which has solution $y(x) = e^x$. The Taylor series of the solution about the initial point $x = 0$ can be constructed directly from the differential equation by using it to compute the derivatives of y and their values at $x = 0$. The next step is to use the Taylor series formula about $x = 0$,

$$y(x) = y(0) + x\, y'(0) + \frac{1}{2!}x^2\, y''(0) + \cdots$$

$$= \sum_{k=0}^{\infty} \frac{x^k}{k!}\, y^{(k)}(0), \quad \text{where } y^{(k)}(x) = \frac{d^k y}{dx^k}. \tag{4.29}$$

Evaluating Equation (4.28) at $x = 0$ and substituting the initial condition gives $y'(0) = y(0) = 1$. Now differentiate Equation (4.28) to obtain

$$\frac{d^2 y}{dx^2} = \frac{dy}{dx}.$$

Setting $x = 0$ and using $y'(0) = 1$ yields $y''(0) = 1$.

In this example the nth derivative $y^{(n)}(x)$ is easily obtained by differentiating Equation (4.28) $n - 1$ times, to give

$$\frac{d^n y}{dx^n} = \frac{d^{n-1} y}{dx^{n-1}}, \quad n = 1, 2, \ldots,$$

and hence $y^{(n)}(0) = 1$ for all n. Thus the Taylor series of the solution expanded about the initial point $x = 0$ is

$$y(x) = \sum_{k=1}^{\infty} \frac{x^k}{k!},$$

which is the Taylor series of e^x, as expected.   ◆

For more complicated differential equations, the computation of the higher derivatives is usually algebraically very involved; however, Maple can normally deal easily with such complications. The next example is chosen to illustrate the method and to show how Maple helps overcome these difficulties.

Example 4.2

Consider the nonlinear equation

$$\frac{d^2y}{dx^2} + xy^2 = 0, \tag{4.30}$$

which is more conveniently cast in the form

$$\frac{d^2y}{dx^2} = -xy^2, \tag{4.31}$$

with initial conditions $y(0) = A$, $y'(0) = B$. Substituting the initial conditions into Equation (4.31) immediately gives $y''(0) = 0$. Now differentiate Equation (4.31) with respect to x to obtain

$$\frac{d^3y}{dx^3} = -y^2 - 2xy\frac{dy}{dx}, \tag{4.32}$$

and on substituting $x = 0$, $y(0) = A$ and $y'(0) = B$ into the right-hand side we obtain $y^{(3)}(0) = -A^2$.   ◆

It should be clear that by repeatedly differentiating Equation (4.32) this method can be extended indefinitely (provided that the derivatives exist, which they clearly do in this case), but that it will become increasingly difficult to keep track of the algebra. Maple makes this task quite easy, but before studying the Maple procedure for this, you should do the following exercise to ensure that you understand the idea.

Exercise 4.23

Differentiate Equation (4.32) twice, to obtain expressions for the fourth and fifth derivatives of $y(x)$. Deduce that $y^{(4)}(0) = -4AB$ and $y^{(5)}(0) = -6B^2$. Then, using Equation (4.29), find the 5th-order Taylor polynomial for the solution of Equation (4.30) (that is, use Equation (4.29) but only for $k = 0, 1, \ldots, 5$).

Before proceeding with a description of the method, we note a rather surprising feature of this type of approximation. Consider the third-order approximation to the solution of Equation (4.30), derived above, that is,

$$y_3(x) = A + Bx - \tfrac{1}{6}A^2x^3.$$

It is good practice, when you have obtained a solution of a differential equation, to substitute it back into the differential equation to check its validity. If we substitute the above approximation into Equation (4.30), we might expect $y_3'' + xy_3^2$ to be proportional to x^4 for small x, that is, $y_3'' + xy_3^2 \simeq cx^4$ for some constant c, because the error term in $y_3(x)$ is of this order. However, since

$$y_3^2 = \left(A + Bx - \tfrac{1}{6}A^2x^3\right)^2 = A^2 + 2ABx + B^2x^2 - \tfrac{1}{3}A^3x^3 + \cdots$$

and $y_3'' = -A^2x$, it turns out that $y_3'' + xy_3^2 = 2ABx^2 + \cdots$. That is, the error when substituting $y_3(x)$ into the equation is of order x^2, which is larger (i.e. grows faster as x increases) than the error in $y_3(x)$ itself. From the above analysis we see that the reason for this is that y_3'' contains no power greater than x, so the x^2 term in xy_3^2 cannot be cancelled. This is a general feature of all second-order equations.

Consequently, if we require an approximate solution for a second-order equation such that the equation is satisfied up to but not including the term in x^N, so that in the present case $y_3'' + xy_3^2 = cx^N + \cdots$ for some constant c, the series must be truncated at, or after, the term in x^{N+1}. This needs to be borne in mind when we substitute in order to check the solution, as above; nevertheless it remains true that, *as an approximation to the true solution*, the error in the approximation $A + Bx - \tfrac{1}{6}A^2x^3$ is proportional to x^4 for small x (that is, it is of fourth order).

The Maple procedure `dsolve` with the `series` option will find the Taylor series solution of Equation (4.30) to a high order with much less effort than that required for Exercise 4.23.

You learned about `dsolve` in *Unit 2*, but not with the `series` option. For more information, type `?dsolve,series` at the prompt.

All that is needed is a set containing the differential equation and the initial conditions. For Equation (4.30) this can be constructed as follows: the equation is

```
>  eq := diff(y(x),x,x)+x*y(x)^2=0:
```

and the initial conditions are

```
>  ic := y(0)=A,D(y)(0)=B:
```

These are used in the command

```
>  sol := dsolve({eq,ic},y(x),series);
```

$$sol := \mathrm{y}(x) = A + Bx - \frac{A^2}{6}x^3 - \frac{AB}{6}x^4 - \frac{B^2}{20}x^5 + O(x^6)$$

The variable named `sol` is an equation with the required Taylor series on the right-hand side, expressed as a series structure. It is important that you understand that what is output here is a *series* structure; in other words it is a Maple `series`. The right-hand side can be extracted and converted into a normal expression using the `rhs` and `convert` commands, to convert the series to a polynomial:

```
>  z := convert(rhs(sol),polynom);
```

$$z := A + Bx - \frac{1}{6}A^2x^3 - \frac{1}{6}ABx^4 - \frac{1}{20}B^2x^5$$

which is the polynomial found in Exercise 4.23.

These commands found the 5th-order Taylor polynomial; i.e. the first ignored term in the series is of order 6. This is because a Maple parameter, `Order`, has its default value set at 6. Higher-order approximations can be obtained by redefining the `Order` parameter before using `dsolve`. Thus the series with x^8 as the highest power is given by setting

Type `?Order` at the prompt for more details. There is also a Maple keyword `order`; however, we shall not be using it.

```
>  Order := 9:
```

and repeating the commands:

```
> sol := dsolve({eq,ic},y(x),series):
> T8 := convert(rhs(sol),polynom);
```

$$T8 := A + B\,x - \frac{1}{6}\,A^2\,x^3 - \frac{1}{6}\,A\,B\,x^4 - \frac{1}{20}\,B^2\,x^5 + \frac{1}{90}\,A^3\,x^6$$
$$+ \frac{1}{63}\,B\,A^2\,x^7 + \frac{13}{1680}\,B^2\,A\,x^8$$

Sometimes the resulting series is not given in its simplest form, and can be expressed more conveniently using `map` together with another command such as `factor`. The following example illustrates this.

Example 4.3

Let us find the Taylor series solution of the equation

$$\frac{d^2y}{dx^2} + x\frac{dy}{dx} + (xy)^2 = 0, \quad y(0) = A, \quad y'(0) = B, \tag{4.33}$$

up to and including the 8th-order term. The equation and initial conditions are defined by typing

```
> eq := diff(y(x),x,x)+x*diff(y(x),x)+(x*y(x))^2=0:
> ic := y(0)=A, D(y)(0)=B:
```

and the required series is found as follows:

```
> Order := 9:  sol := dsolve({eq,ic},y(x),series):
> T8 := convert(rhs(sol),polynom);
```

$$T8 := A + B\,x - \frac{B\,x^3}{6} - \frac{A^2\,x^4}{12} + \left(\frac{1}{40}\,B - \frac{1}{10}\,A\,B\right)x^5 + \left(\frac{A^2}{90} - \frac{B^2}{30}\right)x^6$$
$$+ \left(-\frac{1}{336}\,B + \frac{5}{252}\,A\,B\right)x^7 + \left(-\frac{1}{840}\,A^2 + \frac{1}{105}\,B^2 + \frac{1}{336}\,A^3\right)x^8$$

It is clear by inspection that some coefficients can be factorized, and this is achieved using `map` and `factor`:

```
> T8 := map(factor,T8);
```

$$T8 := A + B\,x - \frac{B\,x^3}{6} - \frac{A^2\,x^4}{12} - \frac{B\,(-1 + 4\,A)\,x^5}{40} + \frac{(A^2 - 3\,B^2)\,x^6}{90}$$
$$+ \frac{B\,(-3 + 20\,A)\,x^7}{1008} + \frac{(-2\,A^2 + 16\,B^2 + 5\,A^3)\,x^8}{1680}$$

Note that here `map` has been applied to an *expression*, rather than to a *list* or to a *set*; you should also try the command `factor(T8)` on the first version of `T8` to see why `map` is necessary.

This is the first time that `map` has been used in this way.

Finally, we can check that this satisfies the original equation by substituting it into Equation (4.33) and simplifying the result by ignoring all powers of x^n and above with, in this case, $n = 8$. The easiest way to do this is with `series` and `simplify`:

```
> test := simplify(series(eval(lhs(eq),y(x)=T8),x=0,8));
```

$$test := \left(-\frac{1}{48}\,B + \frac{17}{90}\,A\,B - \frac{11}{30}\,B\,A^2\right)x^7 + O(x^8)$$

(We have used `lhs` here to extract the left-hand side of `eq`, rather than type it out again. You should also try the `series` command without simplifying it to see why `simplify` is necessary.)

So, as expected, the residual `test` contains only terms of x^{n-1} and higher degree, for $n = 8$.   ◆

Exercise 4.24

(a) Use the method of separation of variables to show that the solution of the differential equation

$$\frac{dy}{dx} = y^2, \quad y(0) = A > 0, \qquad (4.34)$$

is

$$\frac{1}{y} = \frac{1}{A} - x, \quad \text{which becomes} \quad y = \frac{A}{1 - Ax}.$$

If you have studied MST209, you will have encountered this method in Block 1, *Unit 2*.

(b) Write a Maple procedure with a single argument N that computes the Taylor series for the solution of Equation (4.34), using the method described in the text, to any given order N. Show that the 10th-order Taylor series solution found using your Maple procedure is exactly the same as the 10th-order Taylor series of the solution y found in part (a).

(c) With $A = 1$, plot the graphs of the exact solution on $0 \le x \le 2$, restricting the range of y to $[-10, 20]$, together with the Taylor series solutions of orders 5, 10 and 20.

Note that the exact solution is singular at $x = 1/A$, and whilst the function $y = A/(1 - Ax)$ exists for $x > 1/A$, the solution of the differential equation is normally considered to exist only on the interval $x < 1/A$ (due to the initial condition being given at $x = 0$).

Exercise 4.25

Find the 9th-order Taylor series solutions of the following.

(a) $\dfrac{dy}{dx} = \dfrac{\sqrt{y}}{x^3 + y^2}, \quad y(1) = 1$

(b) $\dfrac{d^2y}{dx^2} + x\left(\dfrac{dy}{dx}\right)^2 + (xy)^2 = 0, \quad y(0) = y'(0) = 1$

(c) $\dfrac{d^3y}{dx^3} + \left(\dfrac{d^2y}{dx^2}\right)^2 + y\dfrac{dy}{dx} + \sin(xy) = 0, \quad y(0) = y'(0) = y''(0) = 1$

4.5.2   Range of validity (optional)

The examples above show that Maple allows you to construct high-order Taylor series solutions for a wide variety of differential equations. It is tempting to assume that improved accuracy is obtained by increasing the order of the Taylor series. For x sufficiently close to the initial value this is true, but there is a very important caveat. Usually a Taylor series of a function $f(x)$ about a point $x = a$ is valid only if $|x - a| < r_c$, where the number r_c depends on both the original function $f(x)$ and on a. For $|x - a| > r_c$ the Taylor series does not represent the function, and $(a - r_c, a + r_c)$ is called the **range of validity** of the series expansion. A full explanation of this and methods for estimating r_c are beyond the scope of this course. Suffice to say that when using the Taylor series computed by the method described here, one needs to be careful to ensure that x is in a suitable range. A simple, empirical way of doing this is to compute Taylor polynomials $T_n(x)$ of various orders n and show that, for a given value of x, the differences $|T_{n+1}(x) - T_n(x)|$ decrease as n increases. If the differences do not decrease then it is likely that x lies outside the

If you have studied M203 or M208, you will have seen r_c as the *radius of convergence* and the interval as the *interval of convergence*.

range of validity. This method is not rigorous, and it needs to be used cautiously. However, it does provide a guide to the accuracy of the Taylor series, as illustrated in the following exercise.

Exercise 4.26

(a) Use the `series` command to form the nth-order Taylor series $T_n(x)$ for $n = 3, 4, \ldots, 10$ of the function $f(x) = (1 + x)^{1/4}$ about $x = 0$. Then compute the differences $|T_{n+1}(x) - T_n(x)|$, $n = 3, 4, \ldots, 9$, for $x_0 = 0.5$ and $x_0 = 1.5$.

[*Hint*: Use `T[n]` to represent T_n, and use `unapply` to make `T[n]` a function of `x`.]

`unapply` was introduced in *Unit 2*, Section 2.1.

(b) For this example it can be shown that the Taylor series is valid only if $|x| < 1$, so consider also the differences $|T_{n+1}(x) - T_n(x)|$ for $x = 1.02$ and $x = 0.98$, but with $n = 50, 100, \ldots, 500$; in this case you may need to increase `Digits`.

For most differential equations it is difficult to determine the range of validity of the Taylor series representing the solution. The range of validity can depend on the initial conditions as well as on the form of the differential equation itself, as illustrated in Exercise 4.24. However, one way to estimate the range of validity is to compare the Taylor series solution with a numerical solution, also obtained using `dsolve`. The following exercise shows how this can be done. You may wonder why, if the equations can be solved numerically, a Taylor series is desirable. The reason is that the Taylor series gives a relatively simple formula, explicitly showing the dependence on the initial conditions, and this is often more useful than a numerical solution.

Exercise 4.27

(a) Write a Maple procedure with arguments A, B and N that returns the Taylor series of the solution of the equation

$$y'' + \sin(xy) = 0, \quad y(0) = A, \quad y'(0) = B, \tag{4.35}$$

to a given order N. Show that the coefficient of x^7 in the Taylor series is

$$\frac{1}{504}B - \frac{1}{5040}A^5 + \frac{1}{84}AB^2.$$

(b) Use `dsolve` with the `numeric` option to find the numerical solution of Equation (4.35) in the case $A = B = 1$. Use the methods of Subsection 2.6.3 in *Unit 2* to define a function `fx` representing the numerical solution found by `dsolve`. Then, on a single diagram, plot the graphs of `fx` and the 4th-, 10th- and 21st-order Taylor series solutions for the case $A = B = 1$, on the interval $[-2, 2]$. What do you deduce about the range of validity of the Taylor series solutions in this case?

This use of `dsolve` was described in *Unit 2*.

(c) Plot the graph of `fx` on the interval $[0, 10]$.

4.6 The van der Pol oscillator

The Dutch physicist Balthasar van der Pol (1889–1959) formulated the second-order nonlinear equation

$$\frac{d^2x}{dt^2} + v(x^2 - 1)\frac{dx}{dt} + x = 0, \quad v > 0, \tag{4.36}$$

in order to describe oscillations observed in an electrical circuit. The paper in which this equation was first described was published in 1926, and this described electrical circuits first discussed in 1920. By 1934, when van der Pol published a review, the behaviours of the solutions of this and similar equations were well understood. It is worth noting that this understanding came by deriving formulae which approximate the solutions, and these complement the numerical solutions now generated by computers.

B. van der Pol (1926) 'On relaxation oscillations', *Philosophical Magazine*, **2**, 978–92.

B. van der Pol (1920) 'A theory of the amplitude of free and forced triode vibrations', *Radio Review London*, **1**, 701–10, 754–62.

B. van der Pol (1934) 'The nonlinear theory of electrical oscillations', *Proceedings of the Institute of Radio Engineers*, **22**, 1051–86.

In modern literature, Equation (4.36) is named the **van der Pol equation**. It is important because it is one of the simplest equations to exhibit a particular type of oscillation common to many physical systems. We expand on this remark at the end of this section when examples have been studied.

4.6.1 Solving the van der Pol equation

Van der Pol's equation is nonlinear and admits no solution that can be expressed in terms of a finite number of elementary functions. However, its solutions for all initial conditions behave in a fairly simple manner; in this section you are encouraged to explore and understand these solutions. These explorations will be numerical, but before looking at them in more detail you are asked to work through the following exercise. It shows how the most general van der Pol equation, which has three arbitrary constants, can be reduced to the simple form shown in Equation (4.36).

Exercise 4.28

Equation (4.36) has a single free parameter, v; however, the most general van der Pol equation,

$$\frac{d^2z}{d\tau^2} + \nu(z^2 - a^2)\frac{dz}{d\tau} + \omega^2 z = 0, \quad \nu > 0, \quad a > 0, \quad \omega > 0, \tag{4.37}$$

has three parameters, ν, a and ω. Show that in terms of the rescaled variables x and t, where $z = \alpha x$ and $\tau = \beta t$, the constants α and β may be chosen to convert Equation (4.37) to Equation (4.36) with $v = a^2\nu/\omega$.

Note the distinction between the symbols v and ν, from the Roman and Greek alphabets respectively.

If $v = 0$, the van der Pol equation reduces to the **simple harmonic motion equation** $\ddot{x} + x = 0$, all solutions of which are periodic and can be expressed in the form $x = A\cos(t + \alpha)$. Here the constants A and α are determined by the initial conditions, that is, the values of $x(0)$ and $\dot{x}(0)$. For given values of these initial conditions it can be shown that the solution is unique, and exists for all $t \geq 0$. This remains true in the more general case of $v > 0$.

When $v > 0$, the extra term, $v(x^2 - 1)\dot{x}$, looks like the damping term of the **damped linear harmonic oscillator**

$$\frac{d^2x}{dt^2} + \mu\frac{dx}{dt} + x = 0, \tag{4.38}$$

but with a coefficient $\mu = v(x^2 - 1)$ that depends upon x. If $|x| < 1$ then $\mu < 0$ (since $v > 0$) so the damping is negative. This suggests that the amplitude of the motion will increase. If $|x| > 1$ then $\mu > 0$ so the damping is positive, suggesting that the amplitude of the motion will decrease. Thus we might expect the solution of the van der Pol equation to oscillate.

If you have studied the course MST209 then you will be familiar with the damped linear harmonic oscillator, as well as with simple harmonic motion.

The Maple command `dsolve` with the `numeric` option can be used to find numerical solutions of Equation (4.36) for various initial conditions. An example of a suitable procedure is listed below. To display the solution graphically we have used `odeplot`, which requires the `plots` package, loaded by typing `with(plots)`. We hope that the use of `odeplot` below is fairly self-explanatory, although more information can be obtained by typing `?odeplot` at the prompt.

An alternative method to using `odeplot` was illustrated in the solution to Exercise 4.27(b).

The procedure has four arguments:

ic a list containing the initial conditions $x(0)$ and $\dot{x}(0)$;

v   the numerical value of the parameter v of the van der Pol equation;

T   the final time for which the solution is computed, so that
    T corresponds to T, where $0 \le t \le T$;

N   the number of points used in plotting the graph of the solution
    (this is necessary because the default value, N = 50, sometimes
    produces a jagged graph).

The output of this procedure is the graph of the solution for $0 \le t \le T$, with a title giving the value of v and the initial conditions.

```
>   restart:  with(plots):
>   van01 := proc(ic,v,T,N)
>     local eq,ic1,p,tt:
>     eq := diff(x(t),t,t)+v*(x(t)^2-1)*diff(x(t),t)+x(t)=0;
>     ic1 := x(0)=ic[1], D(x)(0)=ic[2];
>     p := dsolve({eq,ic1},x(t),numeric);
>     # The next line defines the graph title
>     tt := cat("v = ",convert(v,string),
>               ", Initial conditions = ",convert(ic,string));
>     odeplot(p,[t,x(t)],t=0..T,numpoints=N,title=tt);
>   end proc:
```

The title is constructed by the concatenation command `cat`, which takes a sequence of strings as argument and simply joins the strings end-to-end. In order to insert the value of v and the initial conditions into the title, these values are converted into strings via `convert`.

An example of the output from this procedure is shown in Figure 4.6 (overleaf), which was plotted with the command
`van01([5,0],2,50,500);`.

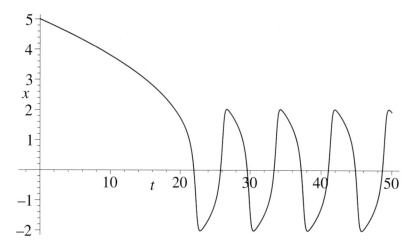

Figure 4.6   Graph of the solution of the van der Pol equation for $v = 2$ and the initial conditions $x(0) = 5$, $\dot{x}(0) = 0$

You should type the procedure `van01` listed above, and check that with the command `van01([5,0],2,50,500);` you can reproduce the graph shown in Figure 4.6. Also, use the value $N = 20$, rather than 500, to see why it is necessary to include this parameter.

Here we see that initially $x(t)$ decreases monotonically from $x(0) = 5$ to $x(t) \simeq 1.7$ at $t = 20$. For $t > 20$ the solution starts to oscillate; for large enough t, these oscillations appear to be regular and the solution appears to be periodic. The first part of the solution, up to the periodic behaviour starting at about $t = 20$, is named the **transient**. As you will see later, the duration of the transient depends on the values of the initial conditions and v.

Exercise 4.29

(a) For initial conditions $x(0) = 5$ and $\dot{x}(0) = 0$, and for $v = 2$, use the `series` option of `dsolve` to show that Equation (4.36) has solution

$$x(t) = 5 - \frac{5}{2}t^2 + 40t^3 - \frac{11\,515}{24}t^4 + \cdots.$$

The use of `dsolve` with `series` was described in Subsection 4.5.1.

(b) On the same diagram and for $0 \leq t \leq 0.1$, plot the fourth-order series solution for $x(t)$ found in part (a) together with the numerical solution obtained using the procedure `van01` given above.

   [*Hint*: As `van01` outputs a graph, use `display` from the `plots` package, as described in *Unit 2*, to produce your diagram.]

4.6.2   Changing the parameter values

You should now investigate the effect of changing the initial conditions and the parameter v. We suggest changing these parameters one at a time, to avoid confusion.

Changing the initial conditions

First set $v = 2$ and consider the effect of changing the initial conditions $x(0)$ and $\dot{x}(0)$.

First fix $\dot{x}(0) = 0$ and vary $x(0)$ from 0.1 to 10 (note that for large $x(0)$ you will need to increase the value of the final time T up to something like 200). What do you notice about the duration of the transient as $x(0)$ increases? Now fix $x(0) = 0$ and vary $\dot{x}(0)$ from 0.1 to 50, all with T set to 40.

From these examples you should observe that for most initial conditions, with $v = 2$, the system eventually settles down to periodic oscillations. However, the duration of the transient increases with increasing $x(0)$ and/or $\dot{x}(0)$. It can be shown that this behaviour is typical for all v and for all initial conditions, except $x(0) = \dot{x}(0) = 0$ (when, as you may have found, $x(t) = 0$ for all $t \geq 0$).

In addition to the individual graphs, you should also plot the graphs of the solutions for all the initial conditions you have considered in a single plot, using the `display` command. Then you should notice that, after the initial transient, the values of $\max(|x(t)|)$ are the same for all initial conditions.

Changing the value of v

Now fix the initial conditions and consider the variation of v.

If $v = 0$, the solution with the initial conditions $x(0) = a$ and $\dot{x}(0) = b$ is

$$x(t) = a \cos t + b \sin t. \tag{4.39}$$

So for all a and b the solution is 2π-periodic, and the amplitude A of the oscillations is given by

$$A = \sqrt{a^2 + b^2}.$$

This clearly depends upon the initial conditions.

If $v > 0$, the solutions behave quite differently. For instance, in the case $v = 2$ considered above, you should have noticed that the amplitude is independent of the initial conditions. You are asked to explore various values of v and T in the following exercise.

Exercise 4.30

Plot the graphs of the solution of van der Pol's equation for the initial conditions $x(0) = 1$ and $\dot{x}(0) = 0$, and the pairs of values of v and T given in the following table.

v	0.01	0.1	0.5	1	2	5	10	20	50
T	500	100	50	50	50	50	50	70	100

From these graphs, using the cursor, estimate the value of $\max(|x(t)|)$ after the initial transients.

(Note that if the value of T is increased too much, then Maple will give the error message 'maxfun limit exceeded'. This can be avoided by using the `dsolve` optional argument, for example `maxfun=100000`. (Alternatively, you can disable `maxfun` with `maxfun=0`.) However, the times suggested in this exercise do not require `maxfun` to be changed.)

`maxfun` limits the number of times the right-hand side of the differential equation is evaluated, in case the user sets parameters that would require the solution to take inconveniently long to compute.

The graphs plotted in Exercise 4.30 show that as v increases, the shape of the oscillations changes. For small v the oscillations of $x(t)$ are approximately sinusoidal, but for large v they have a quite different shape. In Figure 4.7 we show one of these oscillations for the case $v = 100$, $x(0) = -2$ and $\dot{x}(0) = 0$.

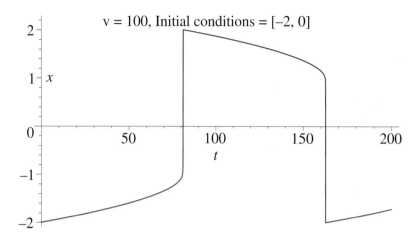

Figure 4.7   Graph of the solution of the van der Pol equation for $v = 100$ and initial conditions $x(0) = -2$, $\dot{x}(0) = 0$

From this graph we see that for most t, $x(t)$ varies relatively slowly. However, at certain regularly spaced times it changes very rapidly from 1 to -2 or from -1 to 2.

This type of oscillation is often named a **relaxation oscillation** and is characterized by the oscillations comprising alternate fast and slow motion, the latter lasting roughly half a period.

4.6.3   The phase plane (optional)

A better picture of the solutions of the van der Pol oscillator, and similar equations, is obtained by plotting the solution in the xy-plane having coordinates x and $y = \dot{x}$, which in this course is named the **phase space** or **phase plane**. A solution curve in the phase plane is called a **phase curve**. The material in this subsection serves partly as an introduction to *Unit 1* of Block B; there you will learn about phase planes in more detail.

If you have studied the course MST209 then you will already have come across some of these ideas.

First note that if $y = \dot{x}$, then Equation (4.36) can be written as the two coupled, first-order equations

$$\frac{dx}{dt} = y, \quad \frac{dy}{dt} = -x - v(x^2 - 1)y. \tag{4.40}$$

Any second-order equation can be written as a pair of first-order equations in this way. It can be shown, for any pair of coupled differential equations of the form $\dot{x} = f(x, y)$ and $\dot{y} = g(x, y)$, where $f(x, y)$ and $g(x, y)$ do not depend explicitly upon t and are sufficiently well behaved, that at most one phase curve passes through any point in the phase plane. This means that a phase curve cannot cross itself, and two phase curves cannot intersect. Further, if the phase curve is a closed loop, the functions $x(t)$ and $y(t) = \dot{x}$ are periodic.

The following code will provide a single phase curve through the point
$(x, y) = (x(0), y(0))$:

```
>  restart:  with(plots):
>  van02 := proc(ic,v,T,N)
>     local eq,ic1,p,tt:
>     eq := diff(x(t),t)=y(t),
            diff(y(t),t)=v*(1-x(t)^2)*y(t)-x(t):
>     ic1 := x(0)=ic[1], y(0)=ic[2]:
>     p := dsolve({eq,ic1},{x(t),y(t)},type=numeric):
>     tt := cat("v = ",convert(v,string),
                ", Initial conditions = ",convert(ic,string)):
>     odeplot(p,[x(t),y(t)],t=0..T,numpoints=N,title=tt):
>  end proc:
```

This procedure was used to draw Figures 4.8 and 4.9, where $v = 2$ and
$v = 10$, respectively; in both cases the initial conditions are
$x(0) = y(0) = 1$, and the time interval is $0 \le t \le 20$.

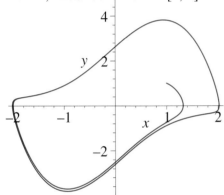

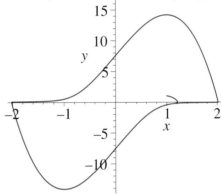

Figure 4.8 Phase curve of the
solution of the van der Pol equation
for $v = 2$ and initial conditions
$x(0) = 1$, $\dot{x}(0) = 1$

Figure 4.9 Phase curve of the
solution of the van der Pol equation
for $v = 10$ and initial conditions
$x(0) = 1$, $\dot{x}(0) = 1$

Exercise 4.31

(a) Use the procedure **van02** to plot the phase curves for the initial
conditions $(x, \dot{x}) = (k/5, 0)$, $k = 1, 2, \ldots, 20$, on a single plot, for $v = 2$,
$T = 20$ and $N = 500$. (One of the easiest ways of doing this is to
create an expression sequence of plot structures using the **seq**
command and then to use the **display** command.)

(b) Repeat part (a) for the initial conditions $(x, \dot{x}) = (0, k/5)$,
$k = 1, 2, \ldots, 20$.

(c) Use the procedure **van02** again, to plot the phase curves with $v = 10$,
$T = 20$ and $N = 2000$, for the initial conditions $(x, \dot{x}) = (k/5, 0)$,
$k = 1, 2, \ldots, 20$ and $(x, \dot{x}) = (0, k)$, $k = 1, 2, \ldots, 15$, respectively.

As a result of these numerical investigations you should be convinced that whatever the value of v and whatever the initial conditions, eventually the solution of the van der Pol equation tends to a closed curve in the phase plane; that is, both $x(t)$ and $y(t)$ tend to a periodic function as $t \to \infty$. In fact, both $x(t)$ and $y(t)$ become indistinguishable from periodic functions quite quickly, provided that their initial values are not too large.

In the following exercise you are asked to investigate the behaviour of the solutions of a different, but similar, equation. This will finish this brief introduction to phase planes.

Exercise 4.32

(a) Consider the differential equation

$$\frac{d^2x}{dt^2} - v\frac{dx}{dt}\cos x + x = 0, \quad v > 0,$$

and, by analogy with Equation (4.36), provide a qualitative description of its solutions as $t \to \infty$.

(b) Check your predictions numerically for various values of v and the initial conditions $(x, \dot{x}) = (a, 0)$ for various $a > 0$, for instance $a = 2, 3, \ldots, 30$ with $v = 1, 2$ and 3.

Summary of Unit 4

This unit consolidates the Maple commands and structures introduced in *Units 1–3*. The two new pieces of Maple introduced here that you should ensure you have a basic working knowledge of are:

- the `series` option of `dsolve`, used to find series solutions of differential equations;
- `odeplot`, used to plot the solutions of differential equations obtained using `dsolve` with the `numeric` option.

Learning outcomes

After studying this unit you should be able to:

- appreciate the use of Maple in a variety of mathematical areas;
- use `dsolve` with the `series` option to find series solutions of differential equations, and also find the series solutions in simple cases 'by hand';
- use `odeplot` to plot numerical solutions of differential equations.

Solutions to Exercises

Solution 4.1

(a) For the sake of definiteness, we have chosen α_1 of the exercise to be $x = \sqrt{3}$; it is easier to type x than alpha1, so there's no harm in making that convenient change. (You will probably have made a different choice.)

Sequence sizes are set to be $N = 1000$. So, use the following Maple commands to generate Spec(x) and Spec($x/(1-x)$) up to N terms:

```
> x := sqrt(3): N := 1000:
> s1 := spec(x,N):
> s2 := spec(x/(x-1),N):
```

Now, join the two expression sequences to form the set $\{s1, s2\}$ and assign it to s as follows:

```
> s := {s1,s2}:
```

Recall that there are N distinct terms in each of the expression sequences s1 and s2, because $x > 1$. If the same term occurs in both s1 and s2, then the number of terms in s will be less than $2N$. This is because the same term does not occur more than once in a set (and Maple respects this property of sets). The command nops is used to count terms in s, so that

```
> nops(s)-2*N;
```
$$0$$

confirms the required result.

(b) Start by defining a Maple function which generates the set of positive integers up to M, $\{1, 2, \ldots, M\}$, using the Maple command

```
> Zp := M->{seq(k,k=1..M)}:
```

Create the expression sequences s1 and s2 as in part (a) for $x = \sqrt{3}$.

Now, we mentioned the minus operation because any positive integers (up to M) that are *not* in the set formed by uniting the spectra will belong to the set defined by Zp(M) minus {s1,s2}. Since $\sqrt{3} < 2$, it will suffice to set M=2*N. Thus, apply the required set minus operation (between Zp(2*N) and {s1,s2}), followed by the Maple function min, which determines the smallest element in the resulting set. This is achieved on a single line as follows:

```
> min(op( Zp(2*N) minus {s1,s2} ));
```
$$1733$$

(Note that op has to be used to convert the resulting set to an expression sequence, because the min function operates on expression sequences and not on Maple sets; see *Unit 2*, Subsection 2.2.2.)

We can infer from this that the smallest integer in Zp(2*N) that is not also in {s1,s2} is 1733, therefore $N_{\max} = 1733$. This value depends on the choice of $x = \sqrt{3}$, with different values being obtained from different choices of x, e.g. for $x = \sqrt{7}$, $N_{\max} = 1609$. Moreover, $N_{\max}$ can be made arbitrarily large (within

the bounds set by Maple) by making N sufficiently large. However, recall that you may need to increase Digits sufficiently beyond its default value in order to cope with these larger values of N. For example, for $N = 50\,000$ you may need Digits:=20.

Solution 4.2

Let us first use Maple to investigate a few cases. For an example where $0 < \alpha < 1$, we will try $\alpha = 1/3$ (called x in the following Maple code):

```
> x := 1/3: y := x/(x+1): N := 50:
> spec(x,N);
```

0, 0, 1, 1, 1, 2, 2, 2, 3, 3, 3, 4, 4, 4, 5, 5, 5, 6, 6, 6, 7, 7, 7, 8, 8, 8, 9, 9, 9, 10, 10, 10, 11, 11, 11, 12, 12, 12, 13, 13, 13, 14, 14, 14, 15, 15, 15, 16, 16, 16

```
> spec(y,N);
```

0, 0, 0, 1, 1, 1, 1, 2, 2, 2, 2, 3, 3, 3, 3, 4, 4, 4, 4, 5, 5, 5, 5, 6, 6, 6, 6, 7, 7, 7, 7, 8, 8, 8, 8, 9, 9, 9, 9, 10, 10, 10, 10, 11, 11, 11, 11, 12, 12, 12

For an example where $\alpha > 1$, we take $\alpha = 4/3$:

```
> x := 4/3: y := x/(x+1): N := 50:
> spec(x,N);
```

1, 2, 4, 5, 6, 8, 9, 10, 12, 13, 14, 16, 17, 18, 20, 21, 22, 24, 25, 26, 28, 29, 30, 32, 33, 34, 36, 37, 38, 40, 41, 42, 44, 45, 46, 48, 49, 50, 52, 53, 54, 56, 57, 58, 60, 61, 62, 64, 65, 66

```
> spec(y,N);
```

0, 1, 1, 2, 2, 3, 4, 4, 5, 5, 6, 6, 7, 8, 8, 9, 9, 10, 10, 11, 12, 12, 13, 13, 14, 14, 15, 16, 16, 17, 17, 18, 18, 19, 20, 20, 21, 21, 22, 22, 23, 24, 24, 25, 25, 26, 26, 27, 28, 28

By trying some more examples, including some irrational $\alpha > 0$, you should notice the following property. Any number occurring n times in Spec(α) will occur $n + 1$ times in Spec($\alpha/(\alpha + 1)$). This also holds for $n = 0$, so that any number missing from Spec(α) will occur just once in Spec($\alpha/(\alpha + 1)$).

Solution 4.3

(a) A suitable program is as follows:

```
>  phi := (1+sqrt(5))/2:
>  N := 1000:
>  a := spec(phi,N):
>  b := spec(phi^2,N):
>  for k from 1 to N do
>     if b[k]<>a[k]+k then
>        print("Failure at k = ",k);
>        break;
>     end if;
>  end do;
>  if k=N+1 then
>     print("Formula fits for all positive
            integers up to k = ",k-1);
>  end if;
```

"Formula fits for all positive integers up to k = ", 1000
Hence Equation (4.4) holds for all $k \leq 1000$.

In fact, by using a different form of `print`, you can avoid the annoyance of inverted commas in the output. In particular, the above print statement can be replaced by `printf("Formula fits for all positive integers up to k = "||(k-1));`.

(b) As suggested, note that
$$\phi^2 = 1 + \phi.$$
So, for all $k \in \mathbb{Z}^+$, we have
$$\begin{aligned} b_k &= \lfloor k\phi^2 \rfloor \\ &= \lfloor k(1+\phi) \rfloor \\ &= \lfloor k + k\phi \rfloor \\ &= k + \lfloor k\phi \rfloor \\ &= k + a_k, \end{aligned}$$
where the penultimate equality follows from noting that, given the definition of floor, $\lfloor k + x \rfloor = k + \lfloor x \rfloor$ for all x and any integer k.

Solution 4.4

For both parts of this exercise, it is convenient to define a Maple function which creates the sequence $\mathrm{Spec}(\alpha; \beta)$ up to the Nth term. This will be called `spec2`, a function of 3 arguments, and is defined as follows:

```
>  spec2 := (x,y,N)->seq(floor(k*x+y),
            k=1..N):
```

We also need to define `Zp` as in Exercise 4.1:

```
>  Zp := M->{seq(k,k=1..M)}:
```

(a) For $m = 2$, two spectra are needed, $S_1 = \mathrm{Spec}(\alpha_1; \beta_1)$ and $S_2 = \mathrm{Spec}(\alpha_2; \beta_2)$, where, from Equation (4.5), we have
$$\alpha_1 = 3, \quad \beta_1 = -1, \quad \alpha_2 = \tfrac{3}{2}, \quad \beta_2 = 0.$$
Now proceed as in Exercise 4.1 by checking that there are no repeated terms in either S_1 or S_2, and that if a number occurs in one spectrum it is not repeated in the other. This is done as follows:

```
>  N := 1000:
>  s1 := spec2(3,-1,N):
>  s2 := spec2(3/2,0,N):
>  s := {s1,s2}:
>  nops(s)-2*N;
```
$$0$$

Then, as in Exercise 4.1(b), use `minus` together with `min` and `op`:

```
>  min(op( Zp(2*N) minus {s1,s2} ));
```
$$1501$$

so $N_{\max} = 1501$.

(b) For $m = 3$, three spectra are needed:
$$S_n = \mathrm{Spec}(\alpha_n; \beta_n), \quad n = 1, 2, 3,$$
where, from Equation (4.5),
$$\begin{aligned} \alpha_1 &= 7, \quad \beta_1 = -3, \\ \alpha_2 &= \tfrac{7}{2}, \quad \beta_2 = -1, \\ \alpha_3 &= \tfrac{7}{4}, \quad \beta_3 = 0. \end{aligned}$$
Proceeding as before:

```
>  N := 1000:
>  s1 := spec2(7,-3,N):
>  s2 := spec2(7/2,-1,N):
>  s3 := spec2(7/4,0,N):
>  s := {s1,s2,s3}:
>  nops(s)-3*N;
```
$$0$$

Finally,

```
>  min(op( Zp(3*N) minus {s1,s2,s3} ));
```
$$1751$$

so $N_{\max} = 1751$.

Solution 4.5

(a) Define a Maple function to be the integrand in Equation (4.6):

```
>  f := x->exp(-x^2)/(x+1):
```

The limits of integration are $a = 1$ and $b = 2$, so Equation (4.8) is given by

```
>  a := 1:  b := 2:  h := (b-a)/3:
>  trap3 := evalf((h/2)*(f(a)+f(b))
            +h*(f(a+h)+f(a+2*h)));
```
$$trap3 := 0.06359098044$$

Hence the value of the definite integral in Equation (4.6) is approximately $0.063\,590\,980\,44$.

(b) Maple gives the following value for the integral, accurate to 10 significant figures:

```
>  evalf(Int(f(x),x=1..2));
```
$$0.05958125008$$

The approximation using 3 trapezia overestimates the value of the integral. The reason can be seen if we plot the integrand and the trapezia, as shown opposite.

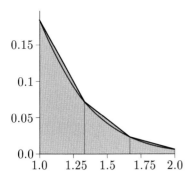

```
> plot([dat]);
```

Clearly the area of the trapezia exceeds the area between the graph of the integrand and the x-axis.

Solution 4.6

For the case of N trapezia, these each have width $h = (b - a)/N$, and their edges occur at $x_0 = a$, $x_1 = a + h$, $x_2 = a + 2h$, ..., $x_N = a + Nh = b$.

The height of the graph at each point x_i is given by $f(x_i)$, so the area enclosed by the trapezia is

$$\frac{h}{2}(f(x_0) + f(x_1)) + \frac{h}{2}(f(x_1) + f(x_2)) +$$

$$\cdots + \frac{h}{2}(f(x_{N-1}) + f(x_N))$$

$$= \frac{h}{2}(f(x_0) + f(x_N)) + h(f(x_1) + f(x_2) +$$

$$\cdots + f(x_{N-1})),$$

which is equivalent to Equation (4.9).

Solution 4.7

(a) Define the procedure as follows:

```
> trap := proc(a,b,N,f)
> local h:
> h := (b-a)/N:
> evalf((h/2)*(f(a)+f(b))+
          h*add(f(a+k*h),k=1..N-1)):
> end proc:
```

(b) Then test the procedure:

```
> f := x->exp(-x^2)/(x+1):
> trap(1,2,3,f);
                0.06359098044
```

which is the same as in Exercise 4.5(a).

(c) First find A:

```
> A := evalf(Int(f(x),x=1..2));
          A := 0.05958125008
```

and then create an expression sequence of the form

$$[3, \log_{10}|A - A_3|],$$
$$[4, \log_{10}|A - A_4|],$$
$$\vdots$$
$$[10, \log_{10}|A - A_{10}|].$$

```
> dat := seq([N,log10(abs(A-trap(1,2,N,f)))],
             N=3..10);
    dat := [3, -2.396884831], [4, -2.646599698], ...
```

(where we have reproduced only the start of the output). Now plot the expression sequence, as follows:

Note that the error $|A - A_N|$ decreases quite rapidly with increasing N, but not quite logarithmically (since the above graph levels out slowly).

Solution 4.8

(a) Since $y = x - a$, $dy = dx$, so changing the integration variable gives

$$\int_a^b f(x)\,dx = \int_0^{b-a} f(y + a)\,dy = \int_0^B g(y)\,dy,$$

where $B = b - a$, and $g(y) = f(y + a)$.

(b) Using the result from part (a), we have

$$\int_a^b f(x)\,dx = \int_0^B g(y)\,dy$$

$$\simeq h \sum_{j=0}^N c_j\, g(jh)$$

$$= h \sum_{j=0}^N c_j\, f(a + jh),$$

where $h = B/N = (b - a)/N$. Hence Equation (4.10) is exact for the same coefficients c_j.

Solution 4.9

Using Equation (4.16):

```
> f := x->exp(-x^2)/(x+1):
> a := 1:  b := 2:  h := b-a:
> evalf(h/2*(f(a)+f(b)));
                0.09502246678
```

Solution 4.10

(a) For $N = 2$, the closed Newton–Cotes formula (Equation (4.11)) becomes

$$\int_0^B g(x)\,dx \simeq \frac{B}{2}\left(c_0\, g(0) + c_1\, g\left(\frac{B}{2}\right) + c_2\, g(B)\right).$$

We require this formula to be exact for all polynomials in $\mathcal{P}_2$, that is, all polynomials of form $g(x) = a_0 + a_1 x + a_2 x^2$. This leads to the equation

$$\int_0^B (a_0 + a_1 x + a_2 x^2)\,dx$$

$$= \frac{B}{2}\left(c_0 a_0 + c_1\left(a_0 + a_1\frac{B}{2} + a_2\frac{B^2}{4}\right)\right.$$

$$\left. + c_2(a_0 + a_1 B + a_2 B^2)\right).$$

Integrating the left-hand side and dividing both sides by $B/2$ gives

$$2a_0 + a_1 B + \tfrac{2}{3}a_2 B^2 = c_0 a_0 + c_1 \left(a_0 + a_1 \frac{B}{2} + a_2 \frac{B^2}{4}\right)$$
$$+ c_2(a_0 + a_1 B + a_2 B^2).$$

Now collect the coefficients of a_0, a_1 and a_2, respectively, on the right-hand side:

$$2a_0 + a_1 B + \tfrac{2}{3}a_2 B^2$$
$$= a_0(c_0 + c_1 + c_2) + a_1 B\left(\frac{c_1}{2} + c_2\right) + a_2 B^2\left(\frac{c_1}{4} + c_2\right).$$

Equating coefficients of a_0, a_1 and a_2 on the two sides:

$$c_0 + c_1 + c_2 = 2, \quad \frac{c_1}{2} + c_2 = 1, \quad \frac{c_1}{4} + c_2 = \tfrac{2}{3}.$$

Solving these equations, either by hand or using Maple, gives

$$c_0 = \tfrac{1}{3}, \quad c_1 = \tfrac{4}{3}, \quad c_2 = \tfrac{1}{3}.$$

Hence the $N = 2$ Newton–Cotes formula is

$$\int_0^B g(x)\,dx \simeq \frac{B}{6}\left(g(0) + 4g\left(\frac{B}{2}\right) + g(B)\right),$$

and, using the result from Exercise 4.8(b), with arbitrary limits of integration, this must give

$$\int_a^b f(x)\,dx \simeq \frac{h}{3}(f(a) + 4f(a + h) + f(b)),$$

where $h = (b - a)/2$.

(b) Using the result from part (a):

```
>   f := x->exp(-x^2)/(x+1):
>   a := 1:  b := 2:  h := (b-a)/2:
>   evalf(h/3*(f(a)+4*f(a+h)+f(b)));
                0.05978061549
```

Solution 4.11

Repeating the code in the text before Exercise 4.11, but with $N = 2$, gives:

```
>   restart:  N := 2:  h := B/N:
>   p := x->add(a[j]*x^j,j=0..N):
>   LH := int(p(x),x=0..B):
>   RH := h*add(c[k]*p(k*h),k=0..N):
>   eq := RH-LH:
>   seqeq := seq(coeff(eq,a[k])=0,k=0..N):
>   seqco := seq(c[k],k=0..N):
>   solve({seqeq},{seqco});
```

$$\{c_2 = \frac{1}{3},\, c_1 = \frac{4}{3},\, c_0 = \frac{1}{3}\}$$

and these values are in agreement with those in Exercise 4.10(a).

Solution 4.12

(a) Gather the sequence of Maple commands starting on page 165 into a procedure:

```
>   NCcoef := proc(N)
>      local h,B,p,x,a,j,LH,RH,c,k,eq,seqeq,
             seqco,sol:
>      h  := B/N:
>      p  := x->add(a[j]*x^j,j=0..N):
>      LH := int(p(x),x=0..B):
>      RH := h*add(c[k]*p(k*h),k=0..N):
>      eq := LH-RH:
>      seqeq := seq(coeff(eq,a[k])=0,k=0..N):
>      seqco := seq(c[k],k=0..N):
>      sol := solve({seqeq},{seqco}):
>      eval([seqco],sol):
>   end proc:
```

Only the last line of the procedure body, namely eval([seqco],sol), is new. This line outputs the values of the coefficients as a list, in the order $c_0, c_1, \ldots, c_N$.

We can check that NCcoef gives the correct values of the coefficients for $N = 1$ and $N = 2$ as follows:

```
>   NCcoef(1);
```

$$\left[\frac{1}{2}, \frac{1}{2}\right]$$

```
>   NCcoef(2);
```

$$\left[\frac{1}{3}, \frac{4}{3}, \frac{1}{3}\right]$$

(b) It is necessary only to add an extra line of Maple code at the beginning (to define B) and a line at the end (to evaluate the right-hand side of Equation (4.10)). We also need to modify the opening lines to include csol as a local variable, and to give NCint the correct inputs. Note that as a is already used in the procedure, we call the integration limits a0 and b0:

```
>   NCint := proc(a0,b0,N,f)
>      local h,B,p,x,a,j,LH,RH,c,k,eq,
             seqeq,seqco,sol,csol:
>      B  := b0-a0:
>      h  := B/N:
>      p  := x->add(a[j]*x^j,j=0..N):
>      LH := int(p(x),x=0..B):
>      RH := h*add(c[k]*p(k*h),k=0..N):
>      eq := LH-RH:
>      seqeq := seq(coeff(eq,a[k])=0,k=0..N):
>      seqco := seq(c[k],k=0..N):
>      sol := solve({seqeq},{seqco}):
>      csol := eval([seqco],sol):
>      evalf(h*sum(csol[k+1]*f(a0+k*h),
             k=0..N)):
>   end proc:
```

Now check that the procedure reproduces the $N = 1$ and $N = 2$ results for the integral in Equation (4.6):

```
> f := x->exp(-x^2)/(x+1):
> NCint(1,2,1,f);
                 0.09502246678
> NCint(1,2,2,f);
                 0.05978061549
```

which are in agreement with the earlier results.

(c) As in the solution to Exercise 4.7(c), we need to create a list of the form

$$[3, \log_{10}|A - A_3|],$$
$$[4, \log_{10}|A - A_4|],$$
$$\vdots$$
$$[10, \log_{10}|A - A_{10}|].$$

```
> A := evalf(Int(f(x),x=1..2));
                 A := 0.05958125008
> dat := seq([N,
              log10(abs(A-NCint(1,2,N,f)))],
              N=3..10);
```

$dat := [3, -4.092617826], [4, -5.107908738], \ldots$

(where we have reproduced only the start of the output). Now plot this:

```
> plot([dat]);
```

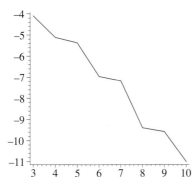

Observe that in this case, even and odd values of N appear to lie on different curves, and that for $N = 10$, the closed Newton–Cotes formula is accurate to around 11 decimal places.

Solution 4.13

A procedure that evaluates the nth-order Taylor series of $\ln(1 + x)$ at $x = 1$ is:

```
> T1 := proc(n)
>    local f,t:
>    f := ln(1+x);
>    t := convert(series(f,x=0,n+1),polynom);
>    evalf(eval(t,x=1));
> end proc:
```

As a check, for $n = 3$ this gives

```
> T1(3);
                 0.8333333333
```

which agrees with $t_3(1) = 1 - \frac{1}{2} + \frac{1}{3} = \frac{5}{6}$.

Now compute the differences `T1(n)-ln(2.0)` in a loop; stop when the magnitude of this difference is less than 0.0063, and print out the value of n:

```
> for n from 3 do
>    d := abs(T1(n)-ln(2.0));
>    if d<0.0063 then print(n);   break
>    end if;
> end do:
                 79
```

So the smallest value of n for which $|t_n(1) - \ln(2)| < 0.0063$ is 79.

Solution 4.14

(a) The function to be approximated is

```
> f := exp(x):
```

and the third-order least squares fit is

```
> N := 3:
> L3 := add(a[k]*x^k,k=0..N);
```

$$L3 := a0 + a1\,x + a2\,x^2 + a3\,x^3$$

The four equations for the four coefficients are found by typing

```
> for k from 0 to N do
>    eq[k] := int(x^k*f,x=0..1)
              =add(a[j]/(k+j+1),j=0..N);
> end do:
```

Now form a set of these four equations and a set of the four variables by typing

```
> eq := {seq(eq[k],k=0..N)}:
> var := {seq(a[k],k=0..N)}:
```

and convert the coefficients of these equations to floating-point numbers:

```
> eq := evalf(eq):
```

Finally, solve the equations by typing

```
> sol := solve(eq,var):
```

and then substitute these values into the approximation L3, truncating the floating-point numbers:

```
> L3 := evalf[6](eval(L3,sol));
```

$$L3 := 0.999060 + 1.01830\,x + 0.421240\,x^2 + 0.278630\,x^3$$

So the third-order least squares fit is indeed that given in the question.

A graph of the difference $\ln(1 + x) - L_3(x)$ is plotted below; this suggests that the largest error occurs at $x = 1$, and is about 0.001.

```
> plot(f-L3,x=0..1);
```

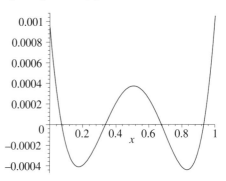

(b) We now compute the difference between the least squares fit and the exact function, and then find the maximum value of this difference. This is done by evaluating the difference at many points in the interval $[0, 1]$, and collecting the largest of these as we proceed. However, the above graph suggests that the largest error is at either $x = 0$ or $x = 1$. We use $M = 100$ points, and the magnitude of the maximum difference is denoted by `maxdiff`, initially set to -1.

```
>   M := 100:  maxdiff := -1:

>   for m from 0 to M do

>       xx := evalf(m/M);

>                      # Compute the difference, d

>       d := abs(exp(xx)-eval(L3,x=xx));

>           # If d > maxdiff then replace maxdiff
>           # by d and keep the value of x where
>           # the maximum difference occurs

>       if d>maxdiff then

>          maxdiff := d:

>          xmax := xx:

>       end if;

>   end do:

>   maxdiff;  xmax;
```

$$0.001051828, 1.$$

These values are consistent with the above graph and, as expected, the largest difference is at $x = 1$ and is 0.0011.

Solution 4.15

(a) If
$$L_N(x) = a_0 + a_1 x + a_2 x^2 + \cdots + a_N x^N,$$
then Equation (4.18) for E becomes
$$E(a_0, a_1, \ldots, a_N)$$
$$= \int_0^1 \left(f(x) - a_0 - a_1 x - a_2 x^2 - \cdots - a_N x^N \right)^2 dx,$$
for $k = 0, 1, \ldots, N$.

Differentiating with respect to a_k gives
$$\frac{\partial E}{\partial a_k} = -2 \int_0^1 x^k (f(x) - a_0 - a_1 x - a_2 x^2 - \cdots - a_N x^N)\, dx,$$
for $k = 0, 1, \ldots, N$. Since $\partial E/\partial a_k = 0$ for all k, this gives
$$\int_0^1 x^k f(x)\, dx$$
$$= \int_0^1 x^k \left(a_0 + a_1 x + a_2 x^2 + \cdots + a_N x^N \right) dx$$
$$= \frac{a_0}{k+1} + \frac{a_1}{k+2} + \cdots + \frac{a_j}{k+j+1} + \cdots + \frac{a_N}{k+N+1}$$
$$= \sum_{j=0}^N \frac{a_j}{k+j+1},$$
for $k = 0, 1, \ldots, N$.

(b) If the interval of interest is $[c, d]$ rather than $[0, 1]$, then all the integrals in part (a) of this solution change from $\int_0^1 \cdots dx$ to $\int_c^d \cdots dx$, so the equations for

the coefficients become
$$\int_c^d x^k f(x)\, dx$$
$$= \int_c^d x^k \left(a_0 + a_1 x + a_2 x^2 + \cdots + a_N x^N \right) dx$$
$$= \sum_{j=0}^N \frac{a_j}{k+j+1} \left(d^{k+j+1} - c^{k+j+1} \right),$$
for $k = 0, 1, \ldots, N$.

Solution 4.16

(a) In this solution we have set `Digits` equal to 15 because the default value is too small (for the 8th-order case) and a value larger than 15 makes a negligible difference. However, you should experiment with different values of `Digits` yourself.

A suitable procedure is as follows. The arguments are: f, the function to be approximated; `interval`, a list [c,d] defining the interval for the approximation; N, the order of the approximation; and `nd`, the number of significant figures used in the final expression.

```
>   restart:

>   Digits := 15:

>   LS := proc(f,interval,N,nd)

>       local LN,a,c,d,k,eq,var,sol:

>       LN := add(a[k]*x^k,k=0..N);

>       c := interval[1];   d := interval[2];

>       for k from 0 to N do

>           eq[k] := int(x^k*f,x=c..d)
>                    =add(a[j]*(d^(k+j+1)
>                    -c^(k+j+1))/(k+j+1),j=0..N);

>       end do;

>       eq := {seq(eq[k],k=0..N)};

>       eq := evalf(eq);

>       var := {seq(a[k],k=0..N)};

>       sol := solve(eq,var);

>       evalf[nd](eval(LN,sol));

>   end proc:
```

Now check this on the function treated in the text, i.e. $f = \ln(1 + x)$, with $N = 2$ on $[0, 1]$:

```
>   f := ln(1+x);
```
$$f := \ln(1 + x)$$
```
>   z := LS(f,[0,1],2,6);
```
$$z := 0.00625953 + 0.915741\, x - 0.233507\, x^2$$

which agrees with the result obtained earlier.

(b) The fourth-order approximations to $\sin x$ and $\cos x$ are found, respectively, by typing

```
>   s := sin(x):  s4 := LS(s,[0,Pi],4,6);
```
$$s4 := 0.00131346 + 0.982601\, x + 0.0544697\, x^2$$
$$- 0.233793\, x^3 + 0.0372093\, x^4$$
```
>   c := cos(x):  c4 := LS(c,[0,Pi],4,6);
```
$$c4 := 0.990963 + 0.0849848\, x - 0.683588\, x^2$$
$$+ 0.145062\, x^3 + 0.312166 \times 10^{-12}\, x^4$$

Now plot a graph of the difference between the exact and approximate functions for $\sin x$, by typing:

```
>  plot(sin(x)-s4,x=0..Pi);
```

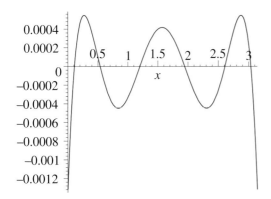

So the maximum difference is about 0.0012.

For $\cos x$, similarly type:

```
>  plot(cos(x)-c4,x=0..Pi);
```

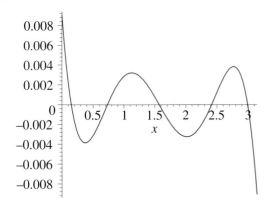

So the maximum difference is just over 0.008.

(c) For the 8th-order approximations we suppress most of the output, to save space. First find the approximations:

```
>  s8 := LS(s,[0,Pi],8,12):
```

```
>  c8 := LS(c,[0,Pi],8,12):
```

Then, for $\sin x$, plot the graph of the difference between $\sin x$ and s8:

```
>  plot(sin(x)-s8,x=0..Pi):
```

We suppress the output here, but the largest differences can be seen to be at $x = 0$ and at $x = \pi$, and these are

```
>  evalf(eval(sin(x)-s8,x=0));
```
$$-0.133443370480 \ 10^{-6}$$

```
>  evalf(eval(sin(x)-s8,x=Pi));
```
$$-0.132767889 \ 10^{-6}$$

So the maximum difference is about 1.3×10^{-7}.

For $\cos x$, typing

```
>  plot(cos(x)-c8,x=0..Pi):
```

shows that again the largest error is at $x = 0$ or $x = \pi$.

From

```
>  evalf(eval(cos(x)-c8,x=0));
```
$$0.1587815 \ 10^{-5}$$

```
>  evalf(eval(cos(x)-c8,x=Pi));
```
$$-0.15790603009596 \ 10^{-5}$$

this maximum error is approximately 1.6×10^{-6}.

Solution 4.17

Let the input to the procedure be a list L and a positive integer n. The first thing to realize is that the N referred to in the question can be calculated as nops(L). So to produce the required output, copy elements of the list L up to and including element $n - 1$, followed by nops(L)+1 and then by the rest of the list. The following uses seq to do the copying:

```
>  insert := proc(L,n)
>     [seq(L[i],i=1..n-1),nops(L)+1,
        seq(L[i],i=n..nops(L))]
>  end proc:
```

Now test the procedure:

```
>  insert([1,2,3],3);
```
$$[1, 2, 4, 3]$$

As this procedure is so simple, it can also be written using the arrow notation for functions, as follows:

```
>  insert := (L,n)->[seq(L[i],i=1..n-1),
        nops(L)+1,seq(L[i],i=n..nops(L))]:
```

```
>  insert([1,2,3],3);
```
$$[1, 2, 4, 3]$$

Solution 4.18

Using a loop within the procedure insertAll, and with insert already defined as in Exercise 4.17, the Maple code is as follows:

```
>  insertAll := proc(L)
>     local i,Lout:
>     Lout := NULL:
>     for i from 1 to nops(L)+1 do
>        Lout := Lout,insert(L,i);
>     end do;
>     Lout;
>  end proc:
>  insertAll([1]);
```
$$[2, 1], [1, 2]$$

```
>  insertAll([1,2]);
```
$$[3, 1, 2], [1, 3, 2], [1, 2, 3]$$

However, once more you should realize that there are many ways of achieving the same result. A different approach, using the methods of *Unit 2*, is as follows:

```
>  insertAll := L->
        seq(insert(L,i),i=1..nops(L)+1):
```

Solution 4.19

The following procedure starts with the solution for $n = 1$ (i.e. $[[1]]$) and then uses a loop to perform `L := map(insertAll,L);` repeatedly until the list of all permutations of n letters is created.

```
>   permute := proc(n)
>     local i,L:
>     L := [[1]];
>     for i from 2 to n do
>       L := map(insertAll,L);
>     end do;
>     L;
>   end proc:
```

Now test the procedure:

```
>   permute(3);
```
$$[[3, 2, 1], [2, 3, 1], [2, 1, 3], [3, 1, 2], [1, 3, 2], [1, 2, 3]]$$

As before, there are plenty of alternative ways to answer this question, and we illustrate some of these here.

For example, the repeated application of `map` can be achieved using the repeated composition operator `@@` as follows. First define a function `step` that applies `map` to its argument:

```
>   step := L->map(insertAll,L):
```

Now the function `permute` can be defined by starting from the solution for $n = 1$ (i.e. $[[1]]$) and then applying the `step` function $n - 1$ times:

```
>   permute := n->(step@@(n-1))([[1]]):
```

(As an aside, the code can be shortened by starting from $n = 0$ rather than $n = 1$. The unique permutation of zero letters is the empty list $[\ ]$, so the function `permute` can be defined as follows:

```
>   permute := n->(step@@n)([[]]):
```

This is surely the shortest solution.)

Another distinct way of achieving the same output is to write a procedure that explicitly expresses the fact that the permutations of n letters are derived from the permutations of $n - 1$ letters. This style of programming is known as **recursive programming**, and the procedure is said to be **recursive**. A simple recursive solution to this problem is as follows:

```
>   permute := proc(n)
>     if n=1 then
>       return [[1]];
>     else
>       return map(insertAll,permute(n-1));
>     end if;
>   end proc:
```

Note that the case $n = 1$ is defined separately, so that the procedure has a place to stop. This style of programming can be very powerful, but you must ensure that every possible input to the procedure eventually terminates. If this does not happen then Maple will go into an infinite series of procedure calls and you will have to interrupt the computational process yourself (using the 'STOP' button on the toolbar). For example, `permute := n->map(insertAll,permute(n-1))` will not terminate.

Solution 4.20

(a) The following procedure meets the specification given in the question:

```
>   misaddressed := proc(L)
>     local i:
>     for i from 1 to nops(L) do
>       if i=L[i] then
>         return false
>       end if
>     end do;
>     true
>   end proc:
```

Now test the procedure:

```
>   misaddressed([2,3,1]);
```
$$true$$
```
>   misaddressed([1,3,2]);
```
$$false$$

(b) The desired permutations can be singled out using `select` as follows:

```
>   select(misaddressed,Lp);
```
$$[[2, 3, 1], [3, 1, 2]]$$

So there are two ways of putting three letters into three envelopes such that no letter is put into the correct envelope.

(c) In order to calculate the number of misaddressed letters for different n, it is convenient to define a Maple function:

```
>   M := n->nops(select(misaddressed,
            permute(n))):
```

Using this function, the $M(n)$ can be calculated for n from 1 to 8:

```
>   seq(M(n),n=1..8);
```
$$0, 1, 2, 9, 44, 265, 1854, 14833$$

Compare this with the expression given in the question:

```
>   seq(n!*add((-1)^i/i!,i=2..n),n=1..8);
```
$$0, 1, 2, 9, 44, 265, 1854, 14833$$

So it is verified that the given expression correctly calculates the number of misaddressed letters for n from 1 to 8. Bernoulli and Euler proved that the expression always gives the correct number of misaddressed letters.

Solution 4.21

(a) There are many ways to construct the desired expression sequence. A simple solution is as follows:

```
>  shuffle := NULL:
>  for i from 1 to 26 do
>     shuffle := shuffle,i,26+i:
>  end do:
>  shuffle;
```

1, 27, 2, 28, 3, 29, 4, 30, 5, 31, 6, 32, 7, 33, 8, 34, 9, 35, 10, 36, 11, 37, 12, 38, 13, 39, 14, 40, 15, 41, 16, 42, 17, 43, 18, 44, 19, 45, 20, 46, 21, 47, 22, 48, 23, 49, 24, 50, 25, 51, 26, 52

(b) Using the hint, the number of shuffles required is computed as follows:

```
>  start := seq(i,i=1..52):
>  p := shuffle:
>  for n from 1 while p<>start do
>     p := seq(p[shuffle[i]],i=1..52);
>  end do:
>  n;
                    8
```

Hence eight perfect shuffles are required to return a pack of 52 cards to its original order. Note the use of the loop counter `n` to count the number of shuffles.

Solution 4.22

Using the hint, we first change the code from Exercise 4.21(a) to create the list resulting from one perfect shuffle.

```
>  makeShuffle := proc(n)
>     local i,shuffle:
>     shuffle := NULL:
>     for i from 1 to n/2 do
>        shuffle := shuffle,i,n/2+i:
>     end do:
>     shuffle;
>  end proc:
```

Check using a small number of cards:

```
>  makeShuffle(4);
                 1, 3, 2, 4
```

This is indeed the result of a perfect shuffle on four cards.

Now we adapt the code from Exercise 4.21(b) to the case of n cards.

```
>  nReturn := proc(n)
>     local p,start,shuffle,k:
>     shuffle := makeShuffle(n);
>     start := seq(i,i=1..n):
>     p := shuffle:
>     for k from 1 while p<>start do
>        p := seq(p[shuffle[i]],i=1..n);
>     end do:
>     k;
>  end proc:
```

As a check we compute the number of shuffles for 52 cards:

```
>  nReturn(52);
                    8
```

The answer computed by this procedure agrees with the answer given in Exercise 4.21.

Now plot the graph over the required range:

```
>  plt := [seq([2*i,nReturn(2*i)],i=1..50)]:
>  plot(plt,style=point,colour=black);
```

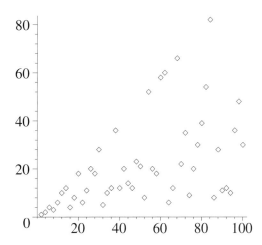

From this plot you can see that for some n the number of perfect shuffles required to return to the original order is small, even when n is large.

Solution 4.23

Differentiate Equation (4.32) with respect to x to obtain

$$\frac{d^4y}{dx^4} = -2y\frac{dy}{dx} - 2\frac{d}{dx}\left(xy\frac{dy}{dx}\right)$$
$$= -2y\frac{dy}{dx} - 2\left(y\frac{dy}{dx} + x\left(\frac{dy}{dx}\right)^2 + xy\frac{d^2y}{dx^2}\right)$$
$$= -4yy' - 2xy'^2 - 2xyy''.$$

Now put $x = 0$, $y(0) = A$, $y'(0) = B$ and $y''(0) = 0$ to obtain $y^{(4)}(0) = -4AB$.

A further differentiation gives

$$\frac{d^5y}{dx^5} = -4yy'' - 4y'^2 - 2(y'^2 + 2xy'y'')$$
$$\quad - 2(yy'' + xy'y'' + xyy''')$$
$$= -6y'^2 - 6yy'' - 6xy'y'' - 2xyy'''.$$

Now make the above substitution together with $y'''(0) = -A^2$ to obtain $y^{(5)}(0) = -6B^2$.

Thus the required 5th-order Taylor polynomial is

$$y(x) = A + Bx - \frac{A^2}{6}x^3 - \frac{AB}{6}x^4 - \frac{B^2}{20}x^5.$$

Solution 4.24

(a) The differential equation $dy/dx = y^2$ is separable, so we can write

$$\int y^{-2}\,dy = \int dx, \quad \text{i.e.} \quad -\frac{1}{y} = x + c.$$

Then $y(0) = A$ gives $c = -1/A$, from which

$$\frac{1}{y} = \frac{1}{A} - x = \frac{1 - Ax}{A},$$

so

$$y = \frac{A}{1 - Ax}.$$

197

(b) A suitable procedure is:

```
> restart:
> T := proc(N)
>    local eq,sol:
>    Order := N+1:
>    eq := diff(y(x),x)=y(x)^2, y(0)=A;
>    sol := dsolve({eq},y(x),series);
>    convert(rhs(sol),polynom);
> end proc:
```

For the 10th-order series solution this gives

```
> z10 := T(10);
```

$$z10 := A + A^2\,x + A^3\,x^2 + A^4\,x^3 + A^5\,x^4 + A^6\,x^5 \\ + A^7\,x^6 + A^8\,x^7 + A^9\,x^8 + A^{10}\,x^9 + A^{11}\,x^{10}$$

The 10th-order Taylor series of the solution found in part (a) is found by typing

```
> t10 := convert(series(A/(1-A*x),x=0,11),
            polynom);
```

$$t10 := A + A^2\,x + A^3\,x^2 + A^4\,x^3 + A^5\,x^4 + A^6\,x^5 \\ + A^7\,x^6 + A^8\,x^7 + A^9\,x^8 + A^{10}\,x^9 + A^{11}\,x^{10}$$

Clearly z10 and t10 are the same, but if in doubt type

```
> d10 := z10-t10;
```

$$d10 := 0$$

(c) Input the formula for the exact solution with $A = 1$:

```
> u := 1/(1-x):
```

and then the various Taylor polynomials, with $A = 1$:

```
> t5 := eval(T(5),A=1):
> t10 := eval(T(10),A=1):
> t20 := eval(T(20),A=1):
```

The required graph is found by typing the following (we show the green and blue plots as grey in the figure).

```
> plot([u,t5,t10,t20],x=0..2,-10..20,
      colour=[black,green,blue,red],
      discont=true);
```

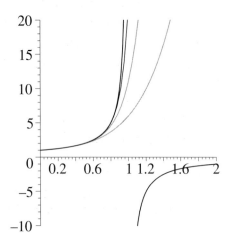

Solution 4.25

The Taylor series are found as follows.

(a)

```
> restart:
> eq :=
    diff(y(x),x)=sqrt(y(x))/(x^3+y(x)^2):
> Order := 10:
> convert(dsolve({eq,y(1)=1},y(x),series),
            polynom);
```

$$y(x) = \frac{1}{2} + \frac{x}{2} - \frac{7\,(x-1)^2}{16} + \frac{11\,(x-1)^3}{32} \\ - \frac{39\,(x-1)^4}{128} + \frac{1697\,(x-1)^5}{5120} - \frac{8247\,(x-1)^6}{20480} \\ + \frac{289291\,(x-1)^7}{573440} - \frac{2944477\,(x-1)^8}{4587520} \\ + \frac{22905251\,(x-1)^9}{27525120}$$

Note how the answer is given as a series of powers of $(x-1)$, because the initial condition is given at $x = 1$.

(b)

```
> restart:
> eq := diff(y(x),x,x)+x*diff(y(x),x)^2
        +x^2*y(x)^2=0:
> Order := 10:
> convert(dsolve({eq,y(0)=1,D(y)(0)=1},y(x),
            series),polynom);
```

$$y(x) = 1 + x - \frac{1}{6}\,x^3 - \frac{1}{12}\,x^4 - \frac{1}{20}\,x^5 - \frac{1}{90}\,x^6 \\ + \frac{1}{72}\,x^7 + \frac{3}{560}\,x^8 - \frac{13}{3240}\,x^9$$

(c)

```
> restart:
> eq := diff(y(x),x$3)+diff(y(x),x,x)^2
        +y(x)*diff(y(x),x)+sin(x*y(x))=0:
> ic := y(0)=1, D(y)(0)=1, (D@@2)(y)(0)=1:
> Order := 10:
> convert(dsolve({eq,ic},y(x),series),
            polynom);
```

$$y(x) = 1 + x + \frac{1}{2}\,x^2 - \frac{1}{3}\,x^3 + \frac{1}{24}\,x^4 - \frac{13}{120}\,x^5 \\ + \frac{1}{18}\,x^6 - \frac{41}{840}\,x^7 + \frac{1619}{40320}\,x^8 - \frac{289}{8064}\,x^9$$

Solution 4.26

(a) First define the function f:

```
> f := (1+x)^(1/4);
```

$$f := (1+x)^{(1/4)}$$

Using the hint given in the question, the required series are computed, as Maple functions of x, within the following loop:

```
>  for n from 3 to 10 do
>     T[n] := unapply(convert(series(
               f,x=0,n+1),polynom),x);
>  end do;
```
(We have suppressed the output here, to save space.)

Now set $x = 0.5$ and form the required differences:
```
>  xx := 0.5:
>  for n from 3 to 9 do
>     abs(T[n+1](xx)-T[n](xx));
>  end do;
```
$$0.002349854$$
$$0.000881195$$
$$0.000348806$$
$$0.000143260$$
$$0.000060438$$
$$0.000026022$$
$$0.000011385$$

It can be seen that these differences decrease with increasing n.

Now set $x = 1.5$ and repeat, producing the following output:
$$0.190338135$$
$$0.214130402$$
$$0.254279852$$
$$0.313309103$$
$$0.396531834$$
$$0.512186952$$
$$0.6722453745$$

These differences show a quite different behaviour and actually increase with n.

(b) For x close to 1 the behaviour is similar:
```
>  for n from 50 to 500 by 50 do
>     m := n+1;
>     T[n] := unapply(convert(series(
               f,x=0,n+1),polynom),x);
>     T[m] := unapply(convert(series(
               f,x=0,m+1),polynom),x);
>  end do:
>  xx := 1.02:  Digits := 20:
>  for n from 50 to 500 by 50 do
>     abs(T[n+1](xx)-T[n](xx));
>  end do;
```
$$0.0041222977849680326$$
$$0.0047156914021269371$$
$$0.0076738110997410533$$
$$0.0144422111568117132$$
$$0.0294427109747680982$$
$$0.0631430309791483247$$
$$0.1402413465055692791$$
$$0.3195684915811234320$$
$$0.7426147075987662052$$
$$1.7525911823029292206$$

```
>  xx := 0.98:
>  for n from 50 to 500 by 50 do
>     abs(T[n+1](xx)-T[n](xx));
>  end do;
```
$$0.0005358712550290755$$
$$0.0000829395498323045$$
$$0.0000182609176437445$$
$$0.46498653889788 \ 10^{-5}$$
$$0.12825658801809 \ 10^{-5}$$
$$0.3721538430331 \ 10^{-6}$$
$$0.1118325821308 \ 10^{-6}$$
$$0.344787433671 \ 10^{-7}$$
$$0.108404227495 \ 10^{-7}$$
$$0.34614534181 \ 10^{-8}$$

Solution 4.27

(a) The following procedure produces the required Taylor series:
```
>  T := proc(N,A,B)
>     local eq,ic,sol:
>     eq := diff(y(x),x,x)+sin(x*y(x))=0;
>     ic := y(0)=A, D(y)(0)=B;
>     Order := N+1:
>     sol := dsolve({eq,ic},y(x),series);
>     convert(rhs(sol),polynom);
>  end proc:
```
and the coefficient of x^7 as given in the question can be verified:
```
>  T7 := T(7,A,B):
>  coeff(T7,x^7);
```
$$\frac{1}{504}B - \frac{1}{5040}A^5 + \frac{1}{84}AB^2$$

(b) First find the numerical solution using `dsolve` by typing
```
>  eq := diff(y(x),x,x)+sin(x*y(x))=0:
>  ic := y(0)=1, D(y)(0)=1:
>  sol := dsolve({eq,ic},numeric):
```
and then, as in *Unit 2*, define a Maple function `fx`:
```
>  fx := x0->eval(y(x),sol(x0)):
```
Then define a function for each of the three Taylor series in turn:
```
>  T4 := unapply(T(4,1,1),x):
>  T10 := unapply(T(10,1,1),x):
>  T21 := unapply(T(21,1,1),x):
```
Finally, plot the graph by typing the following (we show the green and blue plots as grey in the figure).
```
>  plot([fx,T4,T10,T21],-2..2, -2..2.5,
        colour=[red,green,blue,black]);
```

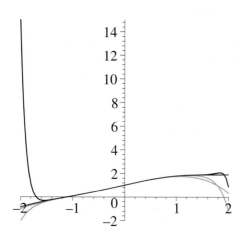

From this figure it would appear that $[-1, 1]$ is a reasonable estimate for the range of validity of the Taylor series, increasing to $[-1.5, 1.5]$ for the 21st-order series.

(c) Continuing with your worksheet from part (b), type

```
> plot(fx,0..10);
```

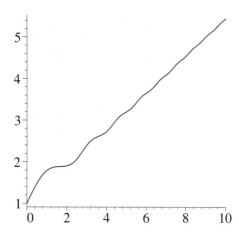

Notice that for large x the solution is almost linear. Can you think of a reason why this is the case?

Solution 4.28

If $z = \alpha x$ and $\tau = \beta t$, then
$$\frac{dz}{d\tau} = \frac{dz}{dx}\frac{dx}{dt}\frac{dt}{d\tau} = \frac{\alpha}{\beta}\frac{dx}{dt}$$
and
$$\frac{d^2z}{d\tau^2} = \frac{d}{d\tau}\left(\frac{dz}{d\tau}\right) = \frac{d}{d\tau}\left(\frac{\alpha}{\beta}\frac{dx}{dt}\right)$$
$$= \frac{d}{dt}\left(\frac{\alpha}{\beta}\frac{dx}{dt}\right)\frac{dt}{d\tau} = \frac{\alpha}{\beta^2}\frac{d^2x}{dt^2},$$
so Equation (4.37) can be written in the form
$$\frac{\alpha}{\beta^2}\ddot{x} + \nu(\alpha^2 x^2 - a^2)\frac{\alpha}{\beta}\dot{x} + \alpha\omega^2 x = 0.$$

On multiplying by β^2/α and reorganizing the second term, this becomes
$$\ddot{x} + \nu a^2\beta\left(\frac{\alpha^2 x^2}{a^2} - 1\right)\dot{x} + \beta^2\omega^2 x = 0.$$

Putting $\alpha = a$, $\beta = 1/\omega$, so $\nu a^2\beta = a^2\nu/\omega = v$, gives Equation (4.36).

Solution 4.29

(a) Typing

```
> eq := diff(x(t),t,t)+v*(x(t)^2-1)
        *diff(x(t),t)+x(t)=0:
> ic := x(0)=5, D(x)(0)=0:
> Order := 5:
> sol := dsolve({eq,ic},x(t),series):
> T3 := convert(rhs(sol),polynom);
```

$$T3 := 5 - \frac{5t^2}{2} + 20vt^3 + \left(-120v^2 + \frac{5}{24}\right)t^4$$

```
> T3 := eval(T3,v=2);
```

$$T3 := 5 - \frac{5}{2}t^2 + 40t^3 - \frac{11515}{24}t^4$$

shows that the fourth-order series solution is as given in the question.

(b) First load the plots package

```
> with(plots):
```

and then proceed as follows:

```
> van01 := proc(ic,v,T,N)
>   local eq,ic1,p,tt:
>   eq := diff(x(t),t,t)+v*(x(t)^2-1)
            *diff(x(t),t)+x(t)=0;
>   ic1 := x(0)=ic[1], D(x)(0)=ic[2];
>   p := dsolve({eq,ic1},x(t),numeric);
>   tt := cat("v = ",convert(v,string),
              ", Initial conditions = ",
              convert(ic,string));
>   odeplot(p,[t,x(t)],t=0..T,numpoints=N,
            title=tt);
> end proc:
```

Typing

```
> display(van01([5,0],2,0.1,50),
          plot(T3,t=0..0.1,colour=black));
```

produces the plot below.

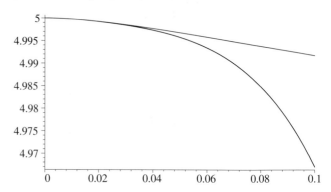

It appears that the Taylor series gives a reasonable approximation to the numerical solution for $0 \le t \le 0.04$.

(If your vertical axis is labelled to a large number of decimal places, then you may need to click on the plot and adjust its width in order to make it look reasonable. Alternatively, include the range x=4.95..5.05.)

Solution 4.30

The procedure as presented in the text is:

```
>   restart:  with(plots):
>   van01 := proc(ic,v,T,N)
>     local eq,ic1,p,tt:
>     eq := diff(x(t),t,t)+v*(x(t)^2-1)
             *diff(x(t),t)+x(t)=0;
>     ic1 := x(0)=ic[1], D(x)(0)=ic[2];
>     p := dsolve({eq,ic1},x(t),numeric);
>     tt := cat("v = ",convert(v,string),
               ", Initial conditions = ",
               convert(ic,string));
>     odeplot(p,[t,x(t)],t=0..T,numpoints=N,
             title=tt);
>   end proc:
```

Here we give the graphs for the first and last pairs of v and T only.

```
>   van01([1,0],0.01,500,500);
```

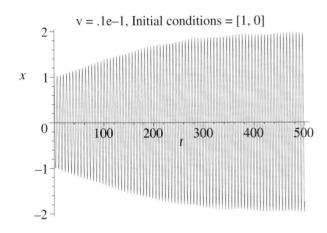

```
>   van01([1,0],50,100,500);
```

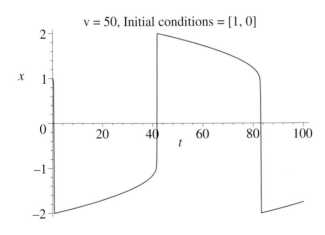

In all cases given in the question you should see that $\max(|x(t)|)$ approaches the value 2, in some cases reaching it very quickly.

A brief discussion of the results of this exercise follows in the text just after the exercise.

Solution 4.31

(a) Each plot is computed separately and combined into an expression sequence, sg, and all are plotted using the display command. Typing

```
>   sg := seq(van02([k/5,0],2.0,20,500),
             k=1..20):
>   display([sg]);
```

produces the following plot.

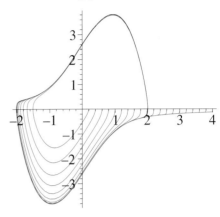

(b) Typing the following produces the required figure (although we do not give the output here):

```
>   sg := seq(van02([0,k/5],2.0,20,500),
             k=1..20):
>   display([sg]);
```

(c) The following pieces of code produce the required output (although, as in part (b), we do not give the output here).

For initial conditions $(x, \dot{x}) = (k/5, 0)$, $k = 1, 2, \ldots, 20$, type

```
>   sg := seq(van02([k/5,0],10.0,20,2000),
             k=1..20):
>   display([sg]);
```

For initial conditions $(x, \dot{x}) = (0, k)$, $k = 1, 2, \ldots, 15$, type

```
>   sg := seq(van02([0,k],10.0,20,2000),
             k=1..15):
>   display([sg]);
```

Solution 4.32

(a) In this example the nonlinear damping term, $-v\dot{x}\cos x$, changes sign when x passes through $(n + \frac{1}{2})\pi$, unlike the van der Pol oscillator where it changes sign only at $x = \pm 1$. This suggests that in the phase space there might be concentric rings of periodic orbits.

(b) A procedure that plots the phase curves of the solution, for given initial conditions and values of v, T and N (the number of points) is as follows (overleaf).

```
>   restart:  with(plots):
>   X := proc(ic,v,T,N)
>      local eq,ic1,p:
>      eq := diff(x(t),t)=y(t), diff(y(t),t)
              -v*y(t)*cos(x(t))+x(t)=0;
>      ic1 := x(0)=ic[1], y(0)=ic[2];
>      p := dsolve({eq,ic1},{x(t),y(t)},
                  type=numeric);
>      odeplot(p,[x(t),y(t)],t=0..T,
              numpoints=N);
>   end proc:
```

Typical solutions for $v = 1.0$, starting at $(x, y) = (12, 0)$ and $(1, 0)$ are found by typing

```
>   a := 12.0:  X([a,0],1.0,20,500);
```

and

```
>   a := 1.0:  X([a,0],1.0,20,500);
```

respectively; the outputs are shown below.

You should be able to see that these tend to different phase curves.

Now plot a number of phase curves with initial conditions $(x, y) = (a, 0)$, with $a = 2, 3, \ldots, 30$ and $v = 2.0$:

```
>   sg := seq(X([a,0],2.0,40,1000),
              a=2..30):
>   display([sg]);
```

The resulting combined plot suggests that there is an infinity of periodic orbits accessed from different initial conditions.

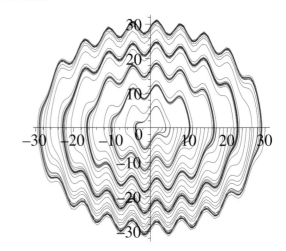

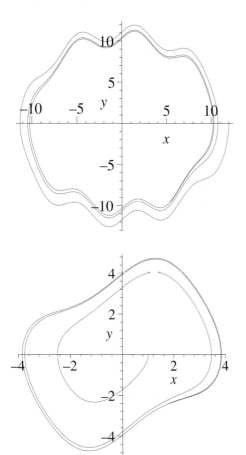

Index